Praise
The Ancestral Mind:
by Gregg D. Ja

"*The Ancestral Mind: Reclaim the Power* is a powerful prescription for improving health and well-being that unites modern science and ancient wisdom. Dr. Jacobs leads us back to a deeper, more fundamental intelligence that we have lost touch with in modern life—the ancestral mind. Reconnecting to this powerful, timeless part of us is one of the most important challenges we face for our collective well-being. Besieged with stress and depression, it us refreshing to learn that we are all endowed with the intrinsic healing capacities of the human brain. This book teaches us how to awaken these capacities to reduce stress, promote healing, cultivate positive emotions, and enhance happiness."

> David S. Sobel, M.D., author of *Healthy Pleasures* and *The Healthy Mind, Healthy Body Handbook*

"*The Ancestral Mind* is clearly an important book for our time, providing tested techniques that enable us to utilize the most fundamental part of ourselves in a powerful way. This book can help anyone find their deeper self and thereby enhance their physical and mental well-being."

> —Andrew Newberg, M.D., coauthor of *Why God Won't Go Away: Brain Science and the Biology of Belief*

"A timely, informative, and refreshing view of the mind. And to access the potential of the ancestral mind, Dr. Jacobs offers the reader effective and easy-to-use techniques."

> —Dr. Ann Webster, instructor in medicine, Harvard Medical School, and staff psychologist, Mind/Body Medical Institute

PENGUIN BOOKS

THE ANCESTRAL MIND

Gregg D. Jacobs, Ph.D., is assistant professor of psychiatry at Harvard Medical School, a senior research scientist at Harvard's Mind/Body Medical Institute, and a research scientist at the Laboratory of Neurophysiology at Harvard Medical School. The author of *Say Good Night to Insomnia*, he lives in Sudbury, Massachusetts, with his wife and two children.

The Ancestral Mind

Reclaim the Power

Gregg D. Jacobs, Ph.D.

PENGUIN BOOKS

PENGUIN BOOKS
Published by the Penguin Group
Penguin Group (USA) Inc., 375 Hudson Street, New York, New York 10014, U.S.A.
Penguin Books Ltd, 80 Strand, London WC2R 0RL, England
Penguin Books Australia Ltd, 250 Camberwell Road, Camberwell, Victoria 3124, Australia
Penguin Books Canada Ltd, 10 Alcorn Avenue, Toronto, Ontario, Canada M4V 3B2
Penguin Books India (P) Ltd, 11 Community Centre,
Panchsheel Park, New Delhi – 110 017, India
Penguin Books (N.Z.) Ltd, Cnr Rosedale and Airborne Roads,
Albany, Auckland, New Zealand
Penguin Books (South Africa) (Pty) Ltd, 24 Sturdee Avenue, Rosebank,
Johannesburg 2196, South Africa

Penguin Books Ltd, Registered Offices:
80 Strand, London WC2R 0RL, England

First published in the United States of America by Viking 2003
Published in Penguin Books 2004

1 3 5 7 9 10 8 6 4 2

THE LIBRARY OF CONGRESS HAS CATALOGED THE HARDCOVER EDITION AS FOLLOWS:
Jacobs, Gregg D.
The ancestral mind : reclaim the power / by Gregg D. Jacobs.
p. cm.
Includes bibliographical references and index.
ISBN 0-670-03217-4 (hc.)
ISBN 0 14 20.0457 X (pbk.)
1. Medicine, Psychosomatic.   2. Mind and body.   3. Mind and body therapies.
4. Stress management.   I. Title.
RC49.J27   2003
616.08—dc21   2003041095

Printed in the United States of America
Set in Palatino
Designed by Erin Benach

To my children, Lauren and Andrew, who are a constant reminder of joy, wonder, and enchantment

In loving memory of my sister Nancy

Contents

The
Ancestral
Mind

PART I

One Body, Two Minds

Introduction

From Whence We Came

If you look at all the things money can buy today, there's no question that we're better off than any generation in history.

In the industrialized world, we're blessed with an abundance of choice in every aspect of life. We have PalmPilots and cell phones and e-mail to help us work more efficiently. We have CD players and wide-screen TVs for wall-to-wall diversion, and a cable channel for every conceivable interest. We have comfortable cars outfitted with all sorts of gadgets, and, theoretically, at least, we could jump on a jet on any day of the week and be anywhere in the world in a matter of hours. We have so much food that we struggle to stay thin. We have safe homes whose climate can be controlled at the touch of a finger. We have a life expectancy a good forty years longer than the average person a century ago.

So why aren't we happy?

With all these toys and gadgets and conveniences, with all these luxuries available to us, why is there so little satisfaction in our day-to-day living? Why do we hear such a litany of complaints from comfortable, middle-class people, all variations on the theme of "I have no time for myself"? Why do so many of us describe our lives as being under "constant pressure" with "too much to do"?

In short, why do we feel so frustrated and so frequently stressed?

As a psychologist working in one of the world's most prominent mind/ body clinics, I can tell you that if you experience such problems they are not "all in your mind." Thirty years after the emergence of mind/body medicine, it's estimated that 75 to 90 percent of all health-care visits still result from stress-related health problems, and that stress is costing American industry a conservatively estimated $150 billion dollars per year in absenteeism, company medical expenses, and decreased productivity.[1,2]

Sleep problems. Digestive disorders. Headaches. Anxiety. Depression. Anger and hostility. Alcoholism and drug abuse. Heart disease. Such stress-related conditions have become epidemic in an affluent, high-tech culture that prides itself on running twenty-four hours a day, seven days a week. Most us have suffered from one or more of these maladies, and for many of us, the symptoms of stress themselves become chronic, and thus another *source* of stress.

The four best-selling drugs in the nation today are for stress-related health problems: ulcer medications, hypertension treatments, tranquilizers and sleeping pills, and antidepressants.[3] As a nation, we spend an astonishing $650 million per year on sleeping pills alone. Four million Americans abuse prescription drugs, and are addicted to tranquilizers, stimulants, or painkillers.[4] One of the great ironies of modern life is that, despite the new global connectedness brought about by the telecommunications revolution, we feel increasingly *disconnected* from ourselves, from others, and from our world. This disconnect is the source of a chronic anxiety. Many of us sense that something is missing in our lives and that, in our hectic existence, focused on getting and spending, on having more and achieving more, we've come to neglect our emotional well-being.

The consequence is an emotional malaise that has undermined our capacity for health and happiness and left us feeling drained as well as confused about how to find meaning. Prozac has become today's vitamin; television today's tranquilizer; and loss of simple joy in life an all-too-common predicament.

■ An Ageless Treatment for Modern Times

I have a prescription for changing this sorry state of affairs. In this book I want to introduce you to an extraordinary, scientifically validated program for improving emotional well-being, reducing illness, easing stress-related symptoms, and countering many of the stress-related causes of death in modern society.

This program is not the type that involves behaviors like eating wisely and exercising conscientiously. It is, rather, an approach that actually *feels good*, and has immediate results. Over time it can bring back the pleasure in living that so many of us have lost. It accomplishes this by enhancing mind/body control, and by producing a mental state that both minimizes unhealthy negative emotions and promotes powerful, life-enhancing positive ones.

If such a device for increasing our inner peace and our sense of well-being were developed today and locked behind an ironclad patent, its inventor could get very rich. But the fact is, the source of these benefits is older than humanity itself. We don't have to adopt something new to improve our lives; what we need is to reengage with a very, very old, very powerful mechanism that has been lost in all the clutter and noise of our modern, technological existence.

And the best part is that this potent antidote to the ravages of stress and enhancement to health and positive emotions actually lies within us. It is a neglected and even disparaged part of ourselves that I call the Ancestral Mind.

We are all familiar with the part of our brains that is the Thinking Mind, the rational, conscious part that processes information, solves problems, and generally helps us make our way through our everyday lives. Western civilization has been built on it, and we have it to thank for most of our material comforts. For all its benefits, however, it is the Thinking Mind (TM, for short) and its products that are also responsible for most of our stresses. Making matters worse, the modern world of commerce is predicated on the belief that the TM is our *only* mind. We've lost sight of the fact that there's another part of ourselves that is accessible to us as a resource for comfort and balance and relief.

The rise to dominance of the Thinking Mind, and the consequent subversion of the Ancestral Mind (AM), is a story as old as human history, but it entered a new phase about four hundred years ago.[5] It was at about this time that the French philosopher René Descartes uttered his famous phrase "I think, therefore I am." The West was entering the Age of Reason, which gave rise to the Industrial Revolution, which in turn was the foundation for the modern world as we know it today.

During the past 400 years of material progress, the Ancestral Mind, a more intuitive entity, and one more at home with feelings and images than with facts and figures, spreadsheets and time cards, has been increasingly relegated to the attic, like some unhinged and embarrassing relation in a gothic novel.

As we'll explore in the chapters that follow, the Ancestral Mind exists just below conscious awareness. It inhabits the brain alongside the Thinking Mind, operating as a separate but related system. The Ancestral Mind is the preverbal part of the brain, the part that guides us through feeling and sensing, and that motivates us to act through the emotions rather than through conscious, rational processes. It relies more on experiential knowledge than on reason. It perceives the "big picture" rather than depending on an intellectual understanding based on a few selected details. It often expresses itself through instincts and intuitions that "put all the pieces together." It is also the reservoir of memories from our own childhood as well as those from our distant collective evolutionary past. As such, it is a source of wisdom and joy, and it provides a solid grounding in times of stress. But most important—and central to the argument of this book—the Ancestral Mind is the part of us that has always been charged with looking out after our fundamental well-being. That was its job through millions of years of the history of humanity, before the Thinking Mind came along to capture the limelight.

Let me assure you, though, that by urging that we reclaim an evolutionarily older part of ourselves, I'm not suggesting that we go back to subsisting on roots and berries and living in grass huts. And rest assured that the Ancestral Mind is not some fuzzy metaphor concocted by a guy with beads and feathers in his hair, romanticizing a version of the past that never existed. The model of the Ancestral Mind that I describe in this book is based on my own research and clinical practice over the past twenty-five years, including fifteen at the Beth Israel Deaconess Medical Center, a major teaching arm of the Harvard Medical School. The concept was developed not at a New Age retreat or a Tibetan monastery, but through a synthesis of recent discoveries in neuroscience, psychology, and mind/body medicine.

What excites me most about this research—and what prompts me to write this book now—is the fact that in just the past few years we have finally assembled sufficient data to present the workings of the Ancestral Mind in powerful and convincing detail. Long shrouded in the murky "data" of folk wisdom, Freudian theories, and anecdotal results, the operations of this hidden part of us can now be observed in the laboratory. We can actually describe the neural pathways that take part and see the functioning of structures within this separate system on a computer screen during magnetic resonance imaging.

Up until now, mind/body medicine had two basic methodologies to offer. The first involved retraining the Thinking Mind to cut off or at least minimize the sort of negative thoughts that induce the "stress response," the constellation of physiological reactions that engulf us when we feel anxious. The other was to use what Dr. Herbert Benson called the "Relaxation Response" in order to short-circuit or at least diminish the bodily effects of that unhealthy response to stress. Both methods are highly effective, and in the second half of this book, I review and explain several techniques based on them in the context of the Ancestral Mind.

But the primary purpose of this book is to enhance that repertoire with novel approaches based on a new level of sophistication in understanding the mind/body connection. This knowledge, which allows us to quiet the TM and reconnect with the Ancestral Mind, shifts the focus from the body to the brain, from conscious to unconscious, and from attempts to ward off negative emotions and illness to the effort to promote positive emotions and health. And it does so in an evolutionary context that makes eminent sense.

In modern times, one of the first individuals to articulate the idea that there was more going on in our minds than what passed through conscious awareness was Sigmund Freud. But his concept of the "unconscious" had a dark preoccupation with repressed urges. The same tightly constrained Victorian world that produced Freud also produced Robert Louis Stevenson's story *Dr. Jekyll and Mr. Hyde*. This novel of two minds inhabiting the same body—one rational and upstanding, the other pulsing with animalistic rage—was representative of the way reason had come to be revered, and of how whatever lay beneath the rational surface—emotions, intuitions, mysteries—had come to be feared. The Ancestral Mind does, in fact, inhabit the same basic real estate as the Thinking Mind, interacts with it, and yet can operate quite independent from it. But it is by no means a raging cesspool of passions that need to be restrained.

The Ancestral Mind is what we see operating when:

• A child plays with a toy in a state of wondrous amazement. She sees the toy for itself and not in terms of its utility, purpose, or relation to anything else. While playing she is unaware of time or sense of self and becomes so completely absorbed in the toy that, to her, it is as alive as she is.

• You are speaking with someone whom you've just met and, although you are not sure why, you take an immediate dislike to him. Through unconscious perception of the person's vocal tone, facial expressions, and body language, you intuitively sense that what he is saying is not what he truly means.

• A hiker stops to watch a pink and orange sunset over snowcapped mountains. The image is so powerful that it commands his attention fully, to the extent of quieting his internal monologue, the constant chatter of the Thinking Mind. In this state of "being" rather than thinking, awareness merges with the present, and the boundaries of self-consciousness disappear. The hiker experiences a feeling of unity, of being a part of something timeless and infinitely greater than the self.

In the chapters that follow, I want to introduce you to the Ancestral Mind in depth: the scientific evidence of its existence, the rationale behind the techniques for accessing it, and a full explanation of the ways it can improve your life by reconnecting you with its power to generate not just better health, but renewed energy, greater concentration, and, ultimately, a more meaningful, joyful life.

During the past three decades of mind/body research, most scientists have focused on the body, not the brain, exploring such phenomena as the hormonal response to stress or psychological states such as anxiety. When I entered graduate school in the early 1980s I was more intrigued by how mind/body techniques exert their therapeutic effects in the brain itself. I had already spent four years working in a hospital's biofeedback laboratory, helping patients to alleviate stress-related health problems such as headaches, gastrointestinal problems, and anxiety. Later, I began a postdoctoral fellowship at the Deaconess Hospital at Harvard Medical School, where I expanded my research to include the mind/body treatment of insomnia. The fellowship included treatment of children and adolescents in the Behavioral Medicine Clinic at Harvard Medical School's Children's Hospital in Boston, where we used biofeedback and relaxation therapies to help children manage a variety of stress-related medical complaints, such as chronic pain, irritable bowel syndrome, colitis, and headaches.

During the years that followed, as I continued to pursue my research and clinical work, the evidence mounted to confirm unconscious emotional processing in the Ancestral Mind, as did evidence of the AM's direct connection to neural pathways that promote healthy mind/body interactions. It became clear to me that a primary evolutionary function of the Ancestral Mind has always been to steer us toward things that elicit positive emotions such as joy. Because such emotions are central to health and well-being, which are in turn fundamental to our very survival, natural selection provided us with an innate drive to experience them. It's our modern preoccupation with the Thinking Mind and its strictures, as well as our alienation from the Ancestral Mind, that blocks this natural propensity.

As we examine the role of these two minds in the world today, we're faced with a painful contradiction. Only a small percentage of humanity takes part in the comfort and security, and the full material benefits made possible by the efforts of the Thinking Mind. And yet we can see that many who do have those benefits are left dissatisfied by the TM's narrow preoccupations and their unintended consequences. Meanwhile, the global economy aims to extend the TM's "benefits" to everyone. Westerners are busily striving to make every culture in every corner of the world run according to the same clock, turning away from ancient traditions of family and village, and embracing the TM's preoccupation with material advancement.

Now, I certainly do not want to suggest that anyone be denied a higher standard of living. My concern, rather, is this: If material advancement means extending the tyranny of the Thinking Mind by removing every last vestige of the Ancestral Mind, where will that leave the human race? Unless we learn to better integrate the two, the future promises little more than billions of stressed and alienated people sitting atop a pile of consumer goods that cannot begin to make them healthy or contented. We must begin to make a concerted investment in the AM and its values, one as committed as that which we have invested in developing the TM.

If we are to improve our lives and see our way toward a more fulfilling future, we must strike a greater balance between thought and emotion, Thinking Mind and Ancestral Mind, as well as establishing an equilibrium between the negative emotions that modern life forces upon us and the positive emotions that are our birthright, and that can help us overcome those negatives. We need to grasp fully the scientific evidence that we are not just creatures of abstract reasoning but also emotional beings who need a more direct experience of life to remain healthy—a far more direct experience than what comes to us filtered through the linear and often anxious Thinking Mind. It is when we temporarily silence our Thinking Mind that we reestablish an essential link with the mental world in which our distant ancestors evolved over millions of years. Like reconnecting with a wise friend that we once knew intimately but from whom we've been cut off, we need to get back in touch with this other, much healthier mental state.

The choice is not an "either/or." The Ancestral Mind is a resource that balances and mitigates the harmful qualities of the Thinking Mind; it doesn't replace it. It *is* possible to live in a high-tech society—we don't seem to have much choice—while also living a greater part of our lives in touch with the healing properties of the Ancestral Mind. But the fact remains that our mind has not caught up with our technology. We now live in a world very different from that which our mind was designed to inhabit—and the conflict is killing us.

By the time you finish this book, you will have gained the knowledge and skills you need to access the Ancestral Mind and to harness its power. I've tried to make the techniques for getting back in touch with this ancient part of ourselves easy to incorporate into your life, putting their benefits well within reach. By reconnecting with the Ancestral Mind, you should be able to achieve a deeper dimension of daily existence that will give you:

- a greater ability to take charge of stress and unhealthy mind/ body interactions
- an improvement in cardiovascular disorders, insomnia, chronic pain, and gastrointestinal problems
- more energy and vibrancy in daily life
- an improved self-awareness, inner strength, and self-esteem.

A by-product of these techniques is that they actually quiet and clear the Thinking Mind, improving the ability to focus and concentrate in daily life. Thus, reconnecting to the Ancestral Mind can actually help us to achieve the fullest potential of *both* minds, improving problem-solving, creativity, performance, and productivity, as well as our general health and well-being. What I hope you will find here is a key to open the door to a new level of consciousness, one that allows you to move toward self-actualization by using your mind's capacities to the fullest.

Chapter One

—■——■—

The Tyranny of the Thinking Mind

The Bible opens with an account of a perfect garden at the beginning of time, where mankind would have dominion, and would be free to partake of everything except the fruit of one particular tree, the tree of the knowledge of good and evil. We all know how that story ends: We bit into the apple and were cast out of Eden. Humanity would go on to advance in many ways, but life would never again be quite as peaceful, and it would certainly no longer be a paradise.

Could it be that "partaking of the tree of knowledge" was really a metaphor for the emergence of the self-conscious Thinking Mind?

Dr. Mihaly Csikszentmihalyi, author of *Flow: Toward a Psychology of Optimal Experience,* conveys the idea of the "paradise" we've lost in precisely these psychological terms:

> The original condition of human beings, prior to the development of self-reflective consciousness, must have been a state of inner peace disturbed only now and again by tides of hunger, sexuality, pain, and danger. Unfulfilled wants, dashed expectations, loneliness, frustration, anxiety, and guilt are likely to have been recent invaders of the mind. They are the dark side of the emergence of consciousness.[1]

Many scholars point to the evolution of language, and a written system for recording it, as the primary catalysts for the development of the self-consciousness that led, perhaps, to our loss of innocence. Language probably evolved around 35,000 years ago, but its written form is only about 8,000 years old.

Language is an essential medium for all the activities we associate with the TM:

- conscious awareness and reflection
- analytical and abstract reasoning
- planning, anticipating, and predicting the future
- problem-solving and skill learning.

We can't go very far along the path of conscious awareness—even to an activity as fundamental as the distinguishing of self from non-self—without some way of defining boundaries and establishing concepts. That's one of the basic functions of words: enabling us to categorize objects and experiences as either known/unknown, recognizable/unrecognizable, useful or irrelevant. In so doing, however, language restricted consciousness by filtering our world through a verbal screen. The dark side of linguistic ability as an aspect of consciousness is that, as soon as something is expressed concretely as a word, the word has already been substituted for the full experience of the thing itself. This loss of immediate experience separated us from the vibrancy of the real world. Instead of being directly in touch with the world, we are only in touch with the words that have come to represent it.

In *The Origins of Consciousness in the Breakdown of the Bicameral Mind*, Julian Jaynes argued that self-consciousness emerged even more recently—at the time of the Bronze Age, some five thousand years ago.[2] According to Jaynes, there was no sense of "I" before this period. It was their first experience of the Thinking Mind's internal monologue, Jaynes speculates, that ancient peoples attributed to hearing the voice of god, or being addressed by spirits.

Once it appeared, the self-conscious Thinking Mind allowed us to radically reshape our environment. Before the advent of the Thinking Mind, the course of human evolution was shaped by natural selection, the mechanism that directs the evolution of every species by weeding out those features that detract from each individual's ability to survive and reproduce in competition with others. The Thinking Mind enabled us to a large extent to step outside natural selection. Through learning and radical innovation we were able to become masters of our fate by dominating nature, but at a great cost: it was this taking control that allowed many "maladaptive" traits, such as persistent anxiety, to emerge—traits that contribute greatly to human illness and unhappiness.

Historian Henri Frankfort characterizes the early condition of precivilized man as one of integral connection with the environment.[3] Our distant ancestors saw and interacted with spirits and gods who took form in every animal, tree, and stone. No distinction was made between subject and object. Because ancient man was too directly immersed in his world to stand outside of it, appearance and reality were not distinct phenomena; dreams and visions were as real as the events of normal life.

In *The Re-Enchantment of Everyday Life,* Thomas Moore describes what he calls the "presiding presence" that once guided us:

> Generations before us have had the sacred, poetic intelligence to speak of angels and demons, and fairies and ghosts, but we have forgotten that wisdom, so taken are we by the allure of facts and figures. Our ancestors knew the world by proper name, but we recognize it only in analytical description. They were so acutely aware of the personality of nature and of things that they could easily give names and faces to things we consider inanimate, and they could even imagine embodied spirits hiding in and near rocks, rivers, mountains, and forests.[4]

This was the context in which the Ancestral Mind evolved, over a period of millions and millions of years.

Once it was able to separate itself from nature, the TM encouraged us to think of ourselves as individuals. There was nothing inherently wrong in that development, but this separation of Self from everything else did mean that we began to *observe* experience rather than *participate* in it. In the process of establishing this "subject-object" distinction, we came to perceive things in terms of their utility and purpose in relation to us, our fears, our past, or our future. We no longer really saw and felt what really existed, because we now viewed things through the distorting veil of words, thoughts, fantasies, and preoccupations, rather than simply as they are.

Once consciousness emerged with full-blown verbal ability, the Thinking Mind was able to assert its dominance over the Ancestral Mind by engaging in an almost continuous internal monologue. If you take a minute or two to close your eyes and simply observe your thoughts, you will probably be amazed at how easily this flood of mental activity wanders continually from the past to the future; hopes to fears; fantasies and desires; arguments and schemes. Eastern meditative traditions have a wonderful name for this tendency—the Monkey Mind—but its effects are very serious.

The internal monologue not only makes us even more self-conscious, but it also alters consciousness by dulling our perceptions of the external world. As we go about our routines, we tune out many of the most positive aspects of life. How many times have you missed a beautiful sunset on the drive home from work because you were fretting over a conversation or event that occurred earlier? We are all too often disengaged from what we are doing: When we cook, we ponder the evening's chores; when we eat, we worry about the phone calls we must return; at times we become so preoccupied with our own thoughts that we do not even hear what is being said to us by someone we love. We have all but forgotten how to "be" in the moment.

As we will explore in later chapters, this relentless internal monologue also plays an important role in causing many of our negative moods and feelings. The endless chatter of the TM adds to the stress that threatens our emotional existence by making us more anxious, hostile, and disconnected.

■ Trapped in Time

Another critical way in which the Thinking Mind's abstract reasoning has distanced us from ourselves is in making time a commodity to be analyzed and abstracted, squeezed and rationed out.

Thoreau once described time as "the stream I go fishing in." No doubt our unique ability to move forward in time mentally and foresee events that have not yet occurred is one of the great benefits of the TM, for it enabled us to anticipate and plan for our future needs. This cognitive time travel, however, came at a price, for too frequently we don't just plan for the future but come to live in it. From our earliest awareness we're taught to believe that what counts most in our lives will occur "when we grow up" or "when we have children" or "when we get that promotion" or "when we retire." We are trained not to seek satisfaction in the present moment but to strive for and expect that happiness to unfold at some future date.

If you objectively observe your thoughts while you are driving, working, playing with your children, or eating dinner, you'll likely be amazed at how much of your mental life involves thoughts about the future. We spend far too large a portion of our lives missing the moment because we keep a never-ending mental list of what will happen in an hour, tomorrow, the coming week, or next year. The Thinking Mind, by constantly pulling us out of the moment, has all but eliminated "now."

Ironically, despite the linear rationality we associate with it, the Thinking Mind's orientation toward the future has made us more fearful than children or even animals. Our ability to project ourselves forward has created new "predators" that are cognitive, rather than biological, threats, and include everything from tomorrow's deadline or next week's court hearing, to even our eventual death. This constant worrying leads to a maladaptive mental state: namely, chronic anxiety.

Preindustrial societies measured time very differently, and time itself had very different meanings. The main divisions of the day were first light, midday, and sunset, and if you were off by an hour or two in your estimation of what constituted "midday," it didn't matter, because nothing happened on that tight a schedule. As late as the nineteenth century Tolstoy could describe Russian peasants showing up to catch a train, unaware of which day it was due to arrive, much less which hour. They simply came to the station, and waited.

In the Western world mechanized timekeeping put an end to that informality and flexibility. By the time of the Middle Ages monks had already begun to use devices like the sundial and the hourglass to mark their schedules for prayers. Later, the tolling of bells in church towers regulated activities for all within hearing. With the development of commercial centers, large clocks on town halls in public squares enabled townspeople to measure time in even more discrete units. When the pocket watch, and later the wristwatch, came along, the dictates of time became personal and internalized. We were all dancing to the same rhythm, no matter where we were.

Time is now our unquestioned master. How many times a day do you look at your watch—fifty, one hundred? Most of what we do—when we arise from bed, go to work, eat lunch, meet with people, get the kids off the bus, watch television shows, go to bed—is based on clock-time. Later in this book, when we explore the scientific literature of sleep, rest, and reverie, we will see that this is not the natural, healthful rhythm the human mind incorporated as it evolved through the eons.

Our unnatural preoccupation with time typically leaves us with a "busy complex," in which we constantly feel that we must be productive rather than allow ourselves to be absorbed in the present. *We're losing time! Time's a-*

wastin'! We have become a culture in which "doing" something, anything, is celebrated. In essence, we're no longer human *beings*, but have become human *doings*.

This goal of keeping busy in order to appear productive is a phenomenon of the last three hundred years—the blinking of an eye, in evolutionary time—and the rise of industrial capitalism. This process became even more pronounced in the second half of the twentieth century. Consider what's happened to the pace of office work in just the last generation. Back when "snail mail" was the standard means of correspondence, you might have to wait a week, or as long as a month, before the post office delivered a response to a letter. When the fax machine entered the picture, you might receive an answer later the same day. With direct service lines on e-mail, you can get a response in "real time"—a reply appears on your screen before you can turn away from your computer.

Such speed can be a real benefit when you want to move ahead with a project as quickly as possible. What's lost in this process, however, is the give-and-take of the natural flow of work. In the past, you dealt with a problem, sent off a document or a proposal, then turned to other tasks while you waited for a response. In the meantime, you might also have lunch with a friend, or take the afternoon off to see your child in a school play.

Now, there's no waiting; the ball is immediately back in your court. This means, of course, that everything demands your prompt attention, even though you have far more things to take care of than could ever be dealt with "at once." Making matters worse, you have no choice but to go along with this pace because your competition—as well as your boss and all your colleagues—are all running just as fast. In practical terms working in "real time" means working ALL the time, because it means that everything has to be done NOW.

E-mail has also begun to encroach on our home lives—not to mention the cell phones and pagers we carry with us everywhere. The situation has become so bad that theaters now have to remind patrons to turn off cell phones before the show begins, much like saloons in the Wild West insisted that patrons check their guns at the door. Restaurants and trains have attempted to set aside special "phone free" areas, but the din of ringing tones and too-loud conversations is still everywhere.

Traveling, even on business, used to be a chance to "get away from the office," to have a breath of fresh air and enjoy a change of scene. Now, thanks to laptops and any number of communication devices, you take the office with you. Which means you never really get away, even to use the time for creative reflection on work. While on the road you are expected not only to do this special "traveling" work—whether meetings or presentations or research—but also to maintain all your usual connections electronically, and thereby do your standard office work as well!

At times it begins to feel as if there's no peace to be found anywhere: Television sets follow you from the airport lounge onto the plane, and even onto the bus that takes you to your rental car. Television programming accosts you in the doctor's waiting room and the auto repair shop. Go shopping, and you're blasted with messages from loudspeakers located above the merchandise. Drive home with your purchases, and the radio exhorts you to go back and buy more.

Think about your day today, which probably involves a mental or written list of activities without a break until you go to bed. We over-schedule ourselves with the demands of work, family, home, and community. Leisure, once a normal part of life, is now a luxury, and yet another scheduled activity. Our society is one in which it no longer suffices to tell a child to "go outside and play." Our children have play dates, lessons, and tightly scheduled events one after the other. In the context of such rigidly organized activities, a parent might now look askance at a child who is simply wandering around the backyard, "playing pretend." Instead, our children now sit for hours in darkened rooms, bombarding their Thinking Minds with electronic stimulation. We're depriving them of the calming companionship of their Ancestral Mind.

The modern world seems determined not merely to compress public time, but to *eliminate* any opportunities for personal time and personal reflection, as if silence and independent thought were subversive activities that needed to be suppressed. And as for having the time and space for quiet awareness of your existence *beyond* the thinking part of ourselves? We've virtually forgotten what that means altogether. In the Western world we've never even formulated a name for this concept, much less bothered to understand the physiology that explains why it is so important for our health.

Before the tyranny of the Thinking Mind took over, we had time to witness the silent drama of a clear night sky, when the mere sight of stars could send us reeling. We had time to respond when the sound of wind rustling through leaves was an invitation to dream, even in the middle of the afternoon. Work was always demanding, but at least it followed the natural rhythms of life. A cycle of planting was followed by one of harvesting, with plenty of feast days in between. Weddings and other celebrations would last for days.[5] And everything was done in close proximity to one's children, one's birthplace, and one's home, with time for recreation and socializing. It wasn't until the eighteenth century that this model changed, and our attitudes about time and labor began the movement toward what they are today. In England, during the early days of industrialization when the modern concepts of work were being invented, villagers had to be forced off the land in order to coerce them into the factories, for no one wanted to cut himself off from the rhythm of life and adapt to the rhythm of a machine. Charged as it is with promoting our well-being, the Ancestral Mind recognized the danger of that way of life.

■ Spend and Acquire

Today, some three hundred years into the industrial (and now *post*industrial) economy, the old adage that "money can't buy happiness" still gets lip service, but the Thinking Mind doesn't seem to be paying attention. More and more, our identities have come to be directly defined by what we own.

Harvard University economist Juliet Schorr, author of *The Overspent American*, has described the phenomenon of the "new consumerism," which equates contentment with acquiring material goods.[6] With advertising and global competition setting the pace, we find ourselves on a treadmill of earn and spend, earn and spend. And because we are spending and acquiring more, we must work harder than ever before. Americans now devote more time to their jobs than any other nation except Japan. Two-income families have become the norm, and one study of married, dual-income couples found that the amount of time involved in working has risen almost seven hours a week over the past twenty years.[7]

It is not uncommon for parents today to work twelve or more hours per day, never seeing their children except on weekends. As a result, America has become, according to Harvard pediatrician T. Berry Brazelton, the least child- and family-oriented society in the world. Dr. Brazelton blames materialism and acquisitiveness, both the provenance of the TM, for parents' not knowing their children and the children not knowing their parents.[8]

In fact, even the power of the old adage about money and happiness may be fading away entirely. When Americans were asked in a national survey what they believed would improve their quality of life, "more money" was the most frequent response, and the more the better. In a survey of a quarter million students entering college, 75 percent of the respondents reported that it was very important or essential that they become very well off financially and that a very important reason for going to college was to make more money. Among nineteen listed objectives, becoming very well off financially placed number one, outranking goals such as developing a meaningful philosophy of life, helping others, and raising a family.[9]

And yet, in a survey of eight hundred college alumni, those who professed similar values—that is, those who preferred a high income and occupational success and prestige to having very close friends and a fulfilling marriage—were twice as likely as their former classmates to describe themselves as "fairly" or "very" unhappy.[10] Recent studies by Dr. Richard Ryan and Dr. Tim Kasser, professors of psychology at the University of Rochester and Knox College, respectively, found that people for whom affluence is a primary focus tend to experience a high degree of anxiety and depression, a lower sense of well-being, and greater behavioral and physical problems.[11] Seduced by the TM's attraction to material trappings, these people have subscribed to distorted values that can actually make them sick. Affluence itself is not the problem, but rather, living a life in which affluence is the focus. Extrinsic goals such as money and material possessions cut us off from elements of the Ancestral Mind that are crucial for well-being, values such as social relationships and the ability to be at peace with ourselves.

Psychologist David Myers and his colleague Dr. Ed Diener, who have studied the relationship between subjective well-being and affluence, summarize the situation this way: Having basic necessities like food and shelter is essential to well-being, but once people are able to afford life's necessities, increasing levels of affluence have a surprisingly weak effect on happiness. In their words, "Wealth, it seems, is like health: Its absence can breed misery, yet having it is no guarantee of happiness."[12] According to Myers, the number of people reporting themselves as very happy has declined slightly over the past forty years.[13] Writing in the journal *American Psychologist*, he notes that we are twice as rich and no happier, while the divorce rate has doubled, teen suicide tripled, reported violence almost quadrupled, and depression rates have soared, particularly among teens and young adults. Compared to their grandparents, today's young adults grow up with much more affluence, slightly less happiness, and much greater risk of depression. Myers terms this conjunction of material prosperity and social recession the "American paradox," and concludes that the modern American dream has become "life, liberty, and the *purchase* of happiness." He describes this fixation as the "greening of America," not in the familiar sense of environmental "green," but in the sense of "greenbacks."

The point of this book is not to question your economic aspirations. But we cannot ignore what is now an established medical fact: the more people deviate from the things that are important to the Ancestral Mind, such as close social ties and altruism, and the more they strive for extrinsic goals of the TM, such as money, the less robust their well-being.

■ Rapid Social Change and Information Overload

Western society, based as it is on the Thinking Mind, thrives on individualism. We exalt the self, in part because individualism, when expressed in the form of "getting ahead," is profitable and powerful. But placing too high a value on individualism can also lead to alienation from others, and even alienation from the self. The premium we place on individualism, and the mythology that has built up around it, makes us willing to uproot ourselves at the drop of a résumé, breaking off relationships with people and places, in pursuit of professional advancement.

Increasingly, the structural underpinnings of our lives are subject to change. In fact, the nature of work and family, and most of our assumptions about gender roles, have been transformed in a single generation. The career-long commitment to a single company, followed by a set pension, is a thing of the past. Our extended families, and sometimes even our nuclear families, are far less integral to our daily existence. And while all but the staunchest conservatives applaud the greater opportunities for women in the workplace and for men in the home that have emerged in the past thirty years, the transition has not been stress-free. For many, the overall rate of change has led to no small degree of confusion, disorientation, and social isolation.

Dr. Barry Schwartz, a psychologist at Swarthmore College, believes that, as well as the disruptions wrought by constant change, the degree of choice in modern life also has become excessive.[14] Anyone who has tried to order telephone service lately would have to agree. Thirty years ago you'd call "the phone company" and choose a plan, and that was that. Now we're all bombarded with solicitations from a dozen different telephone providers, both long distance and mobile. How about ordering coffee in Starbucks? Or a cup of tea almost anywhere? ("We have Darjeeling, English Breakfast, chamomile, Pink Spice, Tutti Frutti, and Secret Sin.") Or a salad in a restaurant? ("Would you like French, Roquefort, Ranch, Creamy Italian, or our own special house dressing . . . dressed or on the side?") Nothing is simple anymore.

We often applaud the fact that religious, ethnic, and gender barriers to mate selection are quickly disappearing, as are gender barriers in the workplace. But as psychologists Robert Woolfolk and Paul Lehrer of Rutgers and Rutgers Medical School point out, along with the array of choices we have in modern life comes an imposing set of responsibilities.[15] For example, we have more freedom than ever before to choose where to live, what career to pursue, whether or not to marry, whether or not to have children, whether to have them early or late, or raise them in traditional or nontraditional arrangements. It is also easier than ever to get married and divorced. The Thinking Mind not only sets up these choices but tempts us with a technological fix should nature not cooperate when, as many couples discover, choosing to put off childbearing means problems with fertility. Couples must choose again—in vitro fertilization, donor egg, any number of other modalities—and yet these remedies don't always work. The psychological anguish this entails, not to mention the expense, has ended many a marriage.

There is no mystery attached to the fact that, in this new era in human history, when for the first time large numbers of people can live unconstrained lives involving high levels of choice, there is a concurrent explosion in depression rates. The burden of responsibility for making innumerable choices can result in a person's becoming psychologically tyrannized by them.

■ The Information Age

We've already mentioned the downside of telecommunications, the media, and the computer in regard to our ownership of time, but they have also disoriented us by skewing our boundaries of space. The ability to have online communication with anyone anywhere in the world has many advantages, as do air travel and the global economy. But it remains true that an almost instantaneous exposure to unfamiliar cultures, people, and ideas requires such rapid adaptation that it is almost inherently stressful. This phenomenon is a variation on jet lag, but takes the form of culture shock.

Even the sheer amount of information bombarding us daily from all corners of the globe is itself enough to overwhelm a person with stress. As Woolfolk and Lehrer note, it was possible as recently as three hundred years ago for one highly learned individual to know everything worth knowing. By the 1940s, it was possible for an individual to know an entire field, such as psychology. Today, the knowledge explosion makes it impossible for one person to master even a significant fraction of one small area of one discipline. Research indicates that information is being generated at such a rapid rate that the amount created in the years 2001 and 2002 alone exceeds all the previous information produced throughout all of human history.[16]

Adding to our stress, the Thinking Mind exposes us to a constant bombardment of negative news. Bad things are an inevitable part of life, but until recently people have not been encouraged to obsess about them as a form of ghoulish entertainment. The media presents us with so much to worry about—murder, catastrophes, terrorism—that we begin to believe that life is inherently horrible. Much of what passes for "news" is irrelevant, sensational, and trauma-driven and would never enter our consciousness were it not for the technology that makes it accessible and the economic incentives behind it.

Consider this sample of the news on a single day as reported by The Boston Globe:

> Mud slide kills three
> Man killed by commuter train
> Woman tortured by her mother
> Robber killed after police chase
> River washes away homes
> Pilot dies in single-engine plane crash
> Two killed in mine blast
> Welder falls to death at construction site
> Dog in critical condition after latest coyote attack

The stress engendered by constant exposure to this type of information is heightened when the media and its crisis mentality panders to fear and outrage. Having expanded the number of outlets and the extent of their coverage, CNN, Fox, and all their competitors in the news business have created a huge expanse of air time that demands to be filled, whether or not there's anything genuinely worth reporting. Their "if it bleeds, it leads" mindset leads to an endless stream of imagery and information whose only purpose is to pull us away from our own lives—which are legitimately the purview of the Ancestral Mind—and deliver our eyeballs to advertisers. The serious side effect here is a significant distortion of reality.

Meanwhile, saturation coverage of such cases as the O. J. Simpson trial or the disappearance of congressional intern Chandra Levy reveal how technology is used to fill our consciousness with facts that, while tragic, have no direct bearing on how we live our daily lives. These "pseudo-events" reduce us to leading lives that are both vicarious and voyeuristic, and almost a parody of the ancient Chinese curse, "May you live in interesting times." These stories aren't inherently interesting, and yet they consume our time on the planet, "amusing us to death," in the words of social critic Neil Postman.

The 1980s brought us a national obsession with abducted children, a problem that, statistically at least, was a cause for alarm about on par with the threat of having a piano fall on one's head. In 1999, as a result of the tragedy at Columbine, shootings in schools become a high-profile media topic. Yet the number of children killed in schools had actually declined for the period 1996–1999 when compared with the previous three years.[17] Likewise, the media's fixation on "road rage" has little basis in fact. Stories reporting an "epidemic" in the number of aggressive drivers are not supported by federal and local transportation records.[18] Government data suggest there is not any statistical or scientific evidence of more aggressive driving; road rage is simply a media invention.

At the same time that television bombards the TM with false impressions of violence, it contributes in a very real way to fostering violent acts. Early in childhood, when young children are developmentally unable to distinguish fact from fiction, television generates deeply skewed impressions about a world where violence is commonplace and is depicted as powerful and exciting. Researchers have shown that infants as young as fourteen months of age incorporate behaviors seen on television, and that young people view an estimated ten thousand acts of violence per year. It should come as no surprise that about one-third of young male felons who committed violent crimes reported consciously imitating crime techniques learned from television.[19]

In the 1970s scientists began to alert the medical community to the fact that television violence increases levels of physical aggressiveness and violence in children.[20] The evidence was so convincing that the American Medical Association's House of Delegates passed a resolution in 1976 stating that "Television violence threatens the health and welfare of young Americans." In the late 1980s additional findings further broadened the consensus.[21] Research in the 1990s strengthened the previously reported links between television violence and aggression in preschoolers, and also found that risky behaviors depicted in the media have been associated with an increase in sexual activity, drinking, smoking, and drug use in both children and adolescents.[22] Sadly, a 2001 survey of research on media violence, sex, and risky behavior that charted the previous decade found that violence today is more graphic and sexual than ever.[23]

In more subtle ways, media imagery also promotes discontent, as it impairs self-esteem and interpersonal relationships. Evolutionary psychologist Dr. David Buss points out that today we are bombarded by images of glamorous and unattainable beautiful people on a scale that has no historical precedent.[24] Buss believes that such imagery may lead to unreasonable expectations about the quantity and quality of sexual partners. He bases this idea on studies which show that men exposed to images of attractive women subsequently rate their commitment to their regular partners as lower.[25] Women exposed to images of high-status men showed a similar decrease in attachment to their regular partners. At the same time, women who are exposed to pictures of other women who are unusually attractive begin to feel less attractive themselves and suffer a subsequent drop in self-esteem; men exposed to images of dominant and influential men suffer comparable reductions in self-concept.[26]

■ Whatever Happened to Joy?

The evidence presented above suggests that the Thinking Mind and all its technological wonders have given rise to an atmosphere that is antithetical to health, that is harmful to the TM itself, and that effectively negates and even obliterates the Ancestral Mind. The TM's mode of abstract analysis is the basis of a worldview that creates boundaries and turns everything into objects; that encourages people to perceive the world as divided into me/not me;

that habitually distinguishes differences, separation, and isolation between self, others, and the environment. Not surprisingly, approaching our lives this way leads to guilt, anxiety, and resentment.

Mystics and philosophers from all religions have observed that man has cut himself off from God to crawl into the confining shell of self-consciousness. Any number of religious traditions hold that self-consciousness and focus on the individual self prevent us from achieving personal growth and attainment of a higher state of awareness through union with our truer, higher self. The fundamental doctrine of so many Eastern religions is that the self-conscious ego is not the real self. The ego can only know itself as an object, and that object "out there" cannot be the self that is truly "in here."

To emerge from this trap, to escape the prison of the personal self, was presented as the highest form of human experience by all great spiritual teachers from Buddha to Jesus. As we will explore in the chapters to come, the hallmark of so many genuinely fulfilling experiences, from mystical ecstasy to states of "flow" in work to romantic love, is loss of the sense of boundaries around the personal self and the feeling of "merging" with something or someone else. This expansion of our sense of self is also known as the experience of joy.

The Thinking Mind keeps us so preoccupied that we seldom experience the simple joy of being alive. In our constant pursuit of success and acquisition, we have lost the ability to just "be." Instead of being absorbed with life, we are absorbed with the TM. As we spend more and more time interacting with machines, controlling nature, and producing more and more things, we feel more isolated from ourselves and from one another.

In a worldview dominated by science and rationality, our need to explain and quantify has itself become irrational. In the process, we have steadily lost alternative modes of being—the experience of wonder, enchantment, and the divine—that are central to health.

Our ancestors felt themselves to be part of a cosmological order that had purpose and meaning. The emergence of the TM with its technology and materialism has fundamentally altered that worldview and left in its place no clear belief system about the meaning of existence. Facts and figures are valuable resources but poor foundations for philosophical convictions. When we seek to explain everything in materialistic terms, we are left with a spiritual void, what humanistic psychologist Erich Fromm called "a malaise, ennui, the automatization of man, the deadening of life."[27] We live, in short, with an existential anxiety brought about by deep and persistent feelings of alienation.

The Ancestral Mind knows better. Predating consciousness, it has no need to ask *why* it is here. It does not strive for explanations, or for "more." It simply is.

Chapter Two

■——■

What *Is* This Ancestral Mind?

Who are you, really?

If you could somehow temporarily shut down your Thinking Mind, if you could strip away your possessions, your work, your worries, plans, and obligations; if you could step out of all the social roles that you are obliged to fill, what would be left?

Would there, in fact, still be a person there? Is there really a "you" that exists apart from those worries, plans, and obligations? Is there really a "you" beyond the cluster of needs, wants, and desires that keeps racing around, trying to keep up with the pace of modern life?

At times, we are able to catch a glimpse of this other, more elemental being inside us. It can reveal itself during a moment of sexual fulfillment or religious awe. It can appear at a moment of intuition in our work, or in the company of the deep resonance we feel when we're miles away from our everyday lives, in some spectacular natural setting. Unfortunately, centuries of dominance by the Thinking Mind make it almost inevitable that we dismiss this glimmer as the product of superstition, or a silly, romantic notion, or perhaps wishful thinking. However, in just the past few years, modern scientific research has followed its own path to a conclusion similar to that reached by many folk and spiritual traditions eons ago: There *is* a deeper part of us that is not about striving or anticipating or having, but simply about *being*. This is the Ancestral Mind, the part of us tied directly to our senses, emotions, and instincts, and the part that used this more grounded awareness to preside over our evolutionary development for millions of years.

In this chapter we're going to review briefly some of this contemporary research. You will be introduced here to a few key brain structures and physiological processes that will help explain what the AM is and how it works, and also how we access it through techniques we'll explore more fully in the chapters to come. In discussing the brain in nontechnical language, we will necessarily have to simplify things. (Ironically, what we're doing, of course, is appealing to your Thinking Mind to convince you of the value and validity of your Ancestral Mind.)

The best way to begin our discussion and get a view of the "big picture" is to follow the path of how our brains came to be precisely what they are.

■ A Very Quick History of the Human Brain

The human brain is the most complex entity in the universe, at least as far as we know. It operates by way of many different structures and systems that overlap and interconnect, often working in parallel, duplicating functions with slight variations. In other instances, what might seem like functions that should be bundled together are actually distributed across different anatomical areas. Adding to the complexity is the fact that the brain is also woven into the rest of the body through an extensive network of nerves and chemical processes.

Part of the reason for our brain's intricacy and mystery is that its current structure developed as the product of innumerable small changes over the course of millions of years. Any change that enhanced our species' survival became part of the regular program, building upon what was already proven successful. This is the nature of evolution through natural selection—it mixes and matches randomly, then goes with what works best.

In working toward a clear understanding of the difference between the Thinking Mind and the Ancestral Mind, and the mechanisms by which we turn down the former to access the latter, we will begin with a quick overview of the three basic physiological levels of the human brain. (For a more detailed overview of the brain, see Appendix A.)

The Reptilian Brain: Sensation and Response

The oldest part of our brain, a structure that came into being long before humanity, is the brain stem, which sits just atop the spine at the base of the skull. Its job is to receive signals from stimuli in the outside world, to initiate movement, and to regulate all the basic life functions—metabolism, heart rate, respiration. The basic reproductive impulse is anchored here, as is territoriality, as are our sensory-motor reactions, such as the startle reflex. This brain stem is called the Reptilian Brain because its basic design and function date back to the Age of Reptiles, roughly 200 million years ago. This part of us is essentially the same brain that operated inside the skull of a crocodile or a lizard eons ago, and still does.

The Reptilian Brain, grounded in the senses, is programmed by genetic instructions. As life evolved into more complex forms, creatures developed more sophisticated and flexible brain structures on top of this basic operating unit; reptiles remained reptiles because they simply kept what they had. The part that we have inherited as the Reptilian Brain still doesn't think, and it does not make emotional connections, any more than the brain of a crocodile does.

In relation to the Ancestral Mind, the critical aspect to remember about the Reptilian Brain is its focus on ensuring basic animal survival by using sensory information to see to it that what takes place inside (our internal physiology) responds to the moment-by-moment requirements of what's outside (our environment, opportunities, and threats). As such, the Reptilian

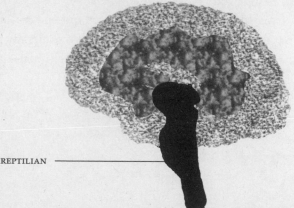

REPTILIAN

The Reptilian Brain

Brain is grounded in the here and now, attending to the details that matter immediately. Although it has a sense of self—an instinctive self—that allows it to navigate the world, it is not self-conscious. It does not stand outside experience and observe it, but rather *is* those experiences.

When sensory signals reach the Reptilian Brain, it makes the necessary adjustments to metabolism, blood pressure, hormone levels, and so on. But in the course of its development, and as part of that same need to adapt to environmental reality in the moment, this primitive brain also developed crude ways of detecting what was going on inside other creatures, often by way of chemical cues that reached it through the sense of smell. The Reptilian Brain can discern such primitive forces as fear and aggression and the desire to mate. These olfactory signals are processed as distinct, "either/or" alternatives, clear options such as "pleasure or pain" or "approach or retreat."

The Reptilian Brain is also the primary location for the large set of physiological reactions that occur instantly in response to any perceived threat— another crude "either/or" split that's come to be known as the stress, or "fight or flight," response. These "instant action" responses, which we'll discuss more fully in a moment, pump us up physically to either fight for our lives or run away.

The Mammalian Brain: Emotional Connections

Further along in evolutionary time, and physically wrapped around the Reptilian Brain, is the more subtle and sophisticated Mammalian Brain. With this

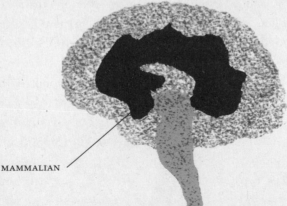

MAMMALIAN

The Mammalian Brain

structure the creatures along the evolutionary road to humanity moved beyond preprogrammed responses and added learning and memory, capabilities still guided by natural selection, but which allow for the fine-tuning of instinctive responses, and thereby, far greater flexibility in behavior.

One of the defining characteristics of mammals is that they give birth to live young, which they then suckle at the breast. Whereas a crocodile will eat her own children, and hatchling crocodiles simply move on independently if allowed, mammalian mothers and infants form attachments, with far greater parental investment in the individual offspring. Although they may not all read *Goodnight Moon* at bedtime, they do establish a bond, which requires that each partner in it be able to register far more information about the other's internal state.

By the mammalian stage in evolutionary history, which commenced about fifty million years ago, what in the Reptilian Brain had been merely crude responses to sensory stimuli—such as fear or sexual arousal—were now taking the first steps on the path to becoming "emotions." When you consider the requirements of an activity like breast-feeding, you can see the value of a more highly developed ability to exchange nonverbal information about internal states. To ensure the offspring's safety, both mother and newborn need to be able to sense the other's presence, to feel comfort when together and stress when separated. It's those comfort/stress signals that compel the pair to stay close, and it's that compulsion that provides the advantage to survival.

So, to the Reptilian Brain's instinctive self, the Mammalian Brain has added the emotional self. In evolutionary terms, given the distance already traveled from the blank, unfeeling stare of a snake to a mare licking her foal to clean it, the rest of the path to our full emotional range is really just a logical extension.[1]

Because more advanced mammals live in packs or troops, with hierarchies of leaders and followers, there are further advantages to be gained from refining the ability to detect "internal states." Rivals and competitors, partners in a hunt, or colleagues in a defensive action all benefit from this exchange of information. Appropriate gestures of dominance and submission, threats and welcomes, were wired into mammalian physiology through "display" behavior, and by the time the more subtle emotions evolved in humans as the language of internal states, pathways had developed that linked this information to signaling through facial expressions.

All the emotions you'd see on *Days of Our Lives* can still be traced back millions of years to this simple system for keeping us in touch with what we need to know in order to survive. With each stage of evolution, the message, the receivers, and the transmitters just kept getting more complex.

It was the most recent stage of human evolution, though, that gave rise to a much different, and often less reliable messaging system entrusted with critical information—cognition.

The Neocortex: Abstract Perceptions

The outer layers of the brain, which enclose the other two more primitive brains, are known as the cortex. This section began to emerge early in the evolutionary history of mammals, increasing in size with the appearance of the primates. The *neocortex*, the six most recently added layers, is vastly larger in humans than in other mammals. While it, too, contributes to our processing of emotional information about the world, its more distinctive purview is planning, sophisticated problem-solving, fine sensory and motor behaviors, and language—all capabilities available to our ancestors tens of thousands of years ago. The neocortex allows us to have ideas, use symbols, and imagine.

It is in the neocortex's most recent contribution to consciousness that we begin to see the paths diverging between Ancestral Mind and Thinking Mind. Only after the emergence of highly specialized neocortical functions

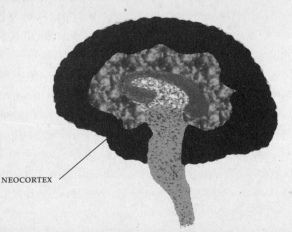

NEOCORTEX

The Neocortex

were we able to *speak* of our emotions and have the "feeling of our feelings." From that basic form of human consciousness emerged the linear, abstract, and rational mental processes that are the foundation of the Thinking Mind, the mind that became self-aware and preoccupied with past and future and began to see itself as the "subject," and everything else as an "object" to be acted upon.

What made this situation so much more complicated—and so much less reliable—was the fact that the Thinking Mind didn't just receive signals from the external world, but was capable of generating signals to create a world of its own, with its own sense of time, space, and self. Whereas the Reptilian Brain is based on pure sensation, and the Mammalian Brain is based on emotion, the neocortex enabled the Thinking Mind to function in response to its own perceptions, concocting its own version of reality by stringing together anticipation, memory, and abstract concepts—all of which contribute to the "internal monologue" that characterizes modern self-consciousness.

■ The Gateway to the Ancestral Mind

In terms of the physiology we have just discussed, the Ancestral Mind does not reside "in" any one level within the brain. If you took an anatomy class, your teacher could not point to a single structure in a diagram and say, "That's it—the seat of the Ancestral Mind." That's because individual brain structures, which gather and process data, have no specific correlations to the phenomena of thought, perceptions, memories, and emotions that we call mind. Rather, they collectively function in a network that ultimately gives rise to the phenomenon of mind.

Yet we know that the Ancestral Mind is real, and that it is a distinct entity, because of what we have learned about particular structures involved in the transmission of signals all the way from the Reptilian Brain to the neocortex. Modern research has elucidated many of the pathways for sensory and emotional responses, and examined how they function with and without the engagement of self-conscious, abstract cognition that is the purview of the Thinking Mind. By consciously altering the neural activity and level of arousal in these structures, we can intentionally experience this wholly different state of mind we call ancestral.

Here are the areas we're going to be targeting as we learn techniques for accessing this calmer, more grounded way of being:

1. The reticular formation: basic alertness

The reticular formation (RF), located in the Reptilian Brain, regulates arousal, attention, stress and relaxation, wakefulness and sleep.[2] The name *reticular* means "network-like," and it does, indeed, send weblike projections throughout the brain. When you hear or see a stimulus that demands some sort of response, it passes through the RF, which notifies other parts of the brain: "Be alert—I'm going to send you important information!" This puts the rest of the system in a heightened state of readiness to respond. Because arousal is fundamental to all mental functions, the RF is a key player in consciousness itself.[3]

2. The thalamus: directing traffic

All sensory signals are processed by the thalamus, located in the "forebrain," atop the brain stem (but for our purposes part of the Mammalian Brain). The thalamus serves as a traffic cop that either opens up and lets the sensory information through, or closes and prevents the information from proceeding any further.[4] It's the thalamus, in conjunction with the RF, that in large part orchestrates the focus and limits of our awareness, determining which sounds, sights, and other sensory messages we consciously perceive from the world around us.

3. The amygdala: setting off the alarm

The amygdala is a pair of almond-shaped structures whose main function is to regulate negative emotional responses like fear and anxiety, and perhaps positive emotions as well.[5] The amygdala is an emotional filter that assesses every situation for potential trouble, assigns emotional meaning to events in our life, and then stores emotional memories that influence our future reactions.

The amygdala also plays a role that is central to traditional mind/body medicine—it sets off the alarm that gives rise to the "fight or flight" response, the tensing for action and revving up of our system as we respond to stress.

The amygdala flips this alarm switch at the slightest provocation. It assesses situations in milliseconds by creating a holistic picture based on just a few pieces of "raw" information, even before the data have been fully analyzed by the neocortex. Existing as it does within the Mammalian Brain, the amygdala does not rely on words or logic, and it certainly does not use restraint.

In prehistoric times the amygdala's remarkable speed was an evolution-ary advantage. Our primitive ancestors walking across the savannahs of East Africa did not have the luxury of pondering and refining the concept of dan-ger. ("Gosh, I wonder if that's a snake. I wonder if it's poisonous. I wonder if it'll try to bite me.") It benefited them to respond immediately and defen-sively, without any thought at all. ("See snake ➤ Jump!") A few false alarms were a small price to pay for the potentially deadly consequences of indeci-sion. What was obviously valuable in the wild, however, can be a serious problem in a modern office, airport, or traffic jam.

4. The hypothalamus: spreading the alarm through the body

Messages from the amygdala ("See snake ➤ Jump!") are passed along to the hypothalamus, situated just under the thalamus. The hypothalamus is re-sponsible for the physical *expression* of emotion—such as when you feel your heart pounding or palms sweating in a stressful situation.[6] It actually carries out the "alert" order sent from the amygdala, causing powerful, physical changes throughout the body, what you experience as the aforementioned "fight or flight" response.

The hypothalamus can set so many reactions in motion because it con-trols the pituitary gland, which regulates the circulation of hormones in the endocrine system. It also regulates the part of the nervous system—the auto-nomic nervous system—that carries out the "automatic" functions of the brain, such as respiration and blood pressure.

A full-body jolt in response to stress includes:

- increased heart rate, respiration rate, and blood pressure for greater physical strength and energy
- increased amounts of stress hormones such as adrenaline to put us "on edge"
- heightened sensory acuity (improved vision and hearing) and faster brain waves for enhanced alertness and mental reactions
- increased muscle tension that might allow us to stand perfectly still or might protect us from injury
- increased sweating to cool the body
- increased blood sugars to reduce fatigue and increase energy
- faster clotting of the blood to minimize blood loss in case all other measures fail

To review: The Ancestral Mind's pathway so far consists of the reticular formation, which says "Heads up! Incoming information!"; the thalamus, which lets the information in or keeps it out; the amygdala, which sets off an alarm if the information seems threatening; and the hypothalamus, which spreads the alarm to jolt the entire body into action.

But then we add the complicating factor of abstract thought.

5. The prefrontal cortex

Located in the forehead, just above the eyes, this specialized area of the cortex is involved in advanced cognitive activities such as reasoning, anticipation, and planning, as well as organizing actions toward a goal. But to achieve these purposes, the PFC must also integrate emotional input. The PFC is where emotion and thought come together, but it is also where the TM's troublesome internal monologue most likely originates.

The PFC works closely with another cortical area that Dr. Andrew Newberg has identified as the "orientation association cortex," which is responsible for locating us in time and space.[7] The orientation cortex synthesizes a constant stream of sensory input to generate a three-dimensional mental representation of the physical boundaries of the self, telling us "this is me, and that is not me," or "I'm here now, and I used to be over there."

Obviously, our sense of being bounded and specifically located plays a significant role in what we call self-consciousness. Another contributor to the self-consciousness of the Thinking Mind is a temporary storage space and processing mechanism within the PFC called "working memory." This space holds representations of those things to which we are currently paying attention, and allows us to go "offline" from the present and bring other thoughts together in abstract fashion to form symbolic representations about the past, present, and future. With it we can merge memories and plans, projecting ourselves backward and forward in time, a facility that intensifies our sense of self in relation to time and place. ("That was me at age six. This will be me at age sixty-six.")

When you come up with a plan for dealing with a problem, solve a puzzle, worry about retirement, or get excited about a party you're going to next week, you are using your PFC and working memory. These functions are also the bases of our ability to self-consciously monitor our own behavior. The PFC, as the meeting ground between emotion and motivation, also acts as a kind of emotional manager that plays a central role in self-control. It contains emotional outbursts by moderating the rapid and often impulsive emotional signals sent from the amygdala, thereby allowing for a more considered and "rational" response to our initial emotional reactions.

Unfortunately, the PFC is also where anxiety-provoking abstract thoughts—"My boss doesn't like me," or "I can't retire on *this* 401(k)!"—can trigger those same "fight or flight" responses, as powerful as the ones that a very real snake crawling over your foot can provoke. It is here that the Thinking Mind begins to create problems for our health, because the amygdala and hypothalamus, which have so much control over our physical reactions to stress, also receive input that originates not in the external (real) world, but from within the perceptions of Thinking Mind.

Clearly, what works so effectively in the face of an immediate, physical threat that can be dealt with by an equally immediate action works less well when the "threats" come in the form of troubling phone messages from your sister, or cryptic e-mails from your boss—problems we may not be able to address anytime soon. Even when the "threat" is not external, or even real, the PFC still feeds information back to the amygdala, which still sounds its alarm, which still prompts the hypothalamus to flood the body with stress hormones, which still triggers a full-fledged fight-or-flight response. The end result: all our energy is revved up, with no outlet for release, which is why stress and anxiety have such a great capacity to detract from our well-being and even make us sick.

Daniel Goleman, author of *Emotional Intelligence*, uses the term "emotional hijacking" to describe emotional "explosions" that occur when the amygdala appraises a situation as an emergency and reacts angrily or even violently, but the PFC cannot modulate its response. According to Goleman, the hallmark of an emotional hijacking is that, once the explosion ultimately passes, we are left with the sensation of not knowing what came over us or why we "blew our fuse."[8]

More commonly, the stress simply builds up and manifests itself in insomnia, headaches, and all the other symptoms associated with spending too much time in our Thinking Mind, and not enough time in the calmer and more grounded environment of our Ancestral Mind.

In more technical terms, the stresses that assault us through the Thinking Mind cause a maladaptive overactivation of the arousal systems in the reticular formation, thalamus, amygdala, and hypothalamus. Because there is no physical release, the stress response throughout the body remains too long in the "on" position. These stresses, along with the tremendous number of stimuli that compete for our attention, in turn cause excessive activation of the association areas of the cortex, which taxes the brain. We waste energy simply processing noise, and the energy drain takes us further out of the present moment.

Our preoccupation with the TM's business—worries, plans, regrets—also undermines our ability to draw upon the positive, healing emotions which are a natural by-product of the Ancestral Mind. The net result is an unhealthy imbalance: too much stress, too much pointless mental activity, too many negative emotions, and not enough soothing, calm, and positive emotions.

That's why we are going to learn techniques in the book not just for turning down the Thinking Mind, but for reanimating the Ancestral Mind as well. We reanimate the Ancestral Mind by reducing arousal in the reticular formation, slowing information flow from the thalamus to the neocortex, and quieting neural activity in the amygdala and higher association cortices. When we close the gates between the thalamus and the cortex, we block the flow to the orientation association cortex (the part that maintains the boundaries between "me" and "not me"), which diminishes our sense of self. As neural activity levels in the brain fall, attention and working memory, both prerequisites for carrying on an internal monologue, decline, resulting in diminished levels of self-consciousness, as well as an altered sense of time. Once the TM is set aside and the Ancestral Mind is engaged, the action downshifts from the cortex and into our Mammalian and Reptilian Brains, which are the realms, not of thought, but of direct sensory and emotional experience.

Many of the techniques we'll learn to reduce arousal and close the gates between the thalamus and the cortex involve repetitive mental stimulation. This kind of stimulation, like the calming rhythm of waves lapping against the beach, induces a kind of cerebral rest by allowing the brain to go "offline" from the activation level that is its energy-demanding default mode, and to conserve rather than expend cortical energy. Acting on the reticular formation and thalamus, these techniques limit the number and types of signals that enter from the outside world. The result is states of quiescence and mental clarity that are themselves pleasurable. And experiencing pleasure in a positive emotion is, itself, part of the healing process of the Ancestral Mind.

The way in which the PFC's structures for self-awareness and goal-directed thought become inactive in the Ancestral Mind is very similar to the process by which we move toward, but not into, sleep. As neural activation levels in the brain decrease, affecting working memory and our normal awareness of self, past, and future, the boundaries of self-consciousness fall away. Lulled by repetitive stimulation, the brain ceases to process the TM's stressful and complex stimuli, and instead of responding as readily to the usual barrage, simply relaxes.

As we mentioned in the introduction, traditional mind/body medicine has offered two ways to escape the tyranny of the Thinking Mind and the stresses it creates:

- Restructuring the TM's harmful cognitive processes themselves
- Learning to temper the fight-or-flight response.

To these we're now adding a third:

- Learning to reengage the Ancestral Mind by consciously altering neural activity in its gateway structures.

But why do we even need such "techniques," no matter how simple and enjoyable? Why can't we just talk ourselves out of all this stress? Why can't we merely "will" ourselves to calm down, relax, and enjoy life more?

We'll turn to those issues next—the fact that the Ancestral Mind not only is overshadowed by the louder and more insistent voice of the Thinking Mind, but also functions outside our conscious awareness. In the next chapter, we'll learn more about the Ancestral Mind's elusive nature and prepare to learn how to overcome that obstacle and consciously reengage with this vital source of our well-being.

Chapter Three

Bypassing the Thinking Mind

Although it is the Thinking Mind of which we are always conscious, it is the Ancestral Mind that looks out for us beneath the level of our awareness. Devoid of rational abstractions, the AM is tied more directly to reality through sensory perception. It speaks to us through the language of emotion, but it is also defined by states of attention: focused attention in the here and now. The AM offers mental stillness that is merged with the present. It offers clarity, imagination, animism, mystery, and direct participation in experience, without the distracting noise of the TM's incessant cerebral chatter. But even this quiescence and absorbed attention circles back to the emotional realm in that such unhurried calm and more direct awareness lead to positive feelings, even feelings of joy or rapture.

Because the Ancestral Mind doesn't use logic and language, it can hardly respond to a well-reasoned argument. It arose long before logic or language appeared, and because it came online first, it has maintained its ability to operate as a separate system, quite independent from the Thinking Mind. The AM learned to communicate emotional information intuitively, nonverbally, and nonconsciously over the course of millions of years, and much of the emotional communication between people today is still nonverbal and unconscious. You can pass someone on the sidewalk or ride with someone on an elevator, and, despite the fact that no words are exchanged, you can intuit a great deal about that person based solely on their body language and facial expression.

The Thinking Mind	The Ancestral Mind
• Verbal, analytical, rational	• Emotional, intuitive, nonverbal
• Consciously aware	• Primarily unconscious
• Abstract	• Physically grounded
• Detached from experience	• Based in experience
• Self-absorbed	• Merely present, with thoughts suspended
• Has a socially conditioned sense of self	• Represents the self that simply is—the instinctual, feeling self
• Bounded by the ego, set apart from others	• Emotionally connected, resonating with others
• Separate from nature	• A part of nature
• Subject/object relations, seeking utility	• Holistic and integrated
• Living in the past and the future	• Here now, in the moment
• Eager to control	• Receptive to emergent realities
• Focused on facts and figures, explanations, cause and effect	• Open to mysteries, childlike, sense of wonder and awe
• Material and having	• Spiritual and "being"
• Sense of time	• Timeless

Or imagine that you're sitting in a meeting with someone you find reasonably attractive, trying to concentrate on the topic at hand. As the conversation continues, you now and again make eye contact. Eventually, though, your gaze lingers a moment too long. Suddenly you recoil and look away. Why did you have to break off the eye contact? Why is it that gazing into someone's eyes feels so . . . intimate? The explanation is that you're experiencing the resonance of two mammalian brains, connected by one of the main sensory channels linking them to the outside world—the optic nerve.

Vocal tone, physical gestures, facial expression, and eye contact transmit underlying emotional messages that words cannot convey—or hide. That's why someone can be polite verbally, yet their emotional signals make us feel uneasy. That instinctive discomfort is the Ancestral Mind at work. It is the

Ancestral Mind, rather than the Thinking Mind, that perceives the unconscious emotional signals that humans give off whether they wish to or not. Our ability to read them is the highly refined outgrowth of the AM's longstanding function of monitoring and adjusting to signals from the outside world—sensory and emotional signals that are essential to our survival.

The Ancestral Mind's gift for nonverbal communication also explains why, for example, we are comforted by the presence of our mammalian pets, and why they give us the feeling that we are not alone. The same ability that enabled mammal mothers and offspring to detect each other's internal states persists today, forging a resonant link between any two mammalian brains. Dogs and cats and horses don't know what we're thinking, but they can and do intuit what we're feeling. The best-selling novel *The Horse Whisperer* was based, in part, on this Ancestral Mind phenomenon of mammalian resonance.

This is also the basis for what we call "animal magnetism" in charismatic people. Individuals who are able to send out and receive information about internal states—that is, emotional signals—better than others are more effective communicators. If, as has been claimed, 90 percent of face-to-face communication is nonverbal, then the most important aspect of forceful communication is not *what* we say but *how* we say it. By making use of the power of nonverbal emotional cues, skillful poker players and dynamic business leaders gain an uncanny sense of what's going on inside other people, information that they put to use for their own gain.

Charismatic religious leaders like Billy Graham or Robert Schuller have that same gift, as do "natural" politicians like Bill Clinton and George W. Bush. Each of them can comfort an audience. In their presence, audiences feel as if their own emotions have resonated and registered. One of the reasons that Al Gore failed in his 2000 presidential bid, of course, was that he lacked this gift of emotional connection to the Ancestral Mind. Mr. Gore strikes most people as being too much a creature of his Thinking Mind, cut off from his primal humanity.

Mind/body techniques that quiet the TM, which we'll discuss in later chapters, help minimize our own internal monologue, our self-conscious awkwardness, our fears and anxieties. Freed of such burdens, we are better able to communicate with ourselves and others. These techniques clear the channels of static, freeing the AM to really see and hear what exists beneath the superficial level of conscious abstraction.

Just because we can't always articulate reasons for our reactions to such signals doesn't mean that we are not receiving them. Of all the information that flows into our brain from our sensory systems every second, only a fraction enters our conscious awareness. As we have seen, the reticular formation and thalamus work in concert as filters to keep the sensory input that we consciously process to a manageable level.

Consider the information your brain is receiving at this very moment: the content of this book, the feel of the chair you may be sitting in, bodily sensations like your heart beating or tension in your neck, sounds around you, and so on. Although your brain is processing all of this data, you are aware of, at most, a tiny fraction of it.

If we were consciously aware of all the stimuli that reached the brain, we would be overwhelmed. In the wisdom of the Ancestral Mind, the reticular formation and thalamus protect us by using an innate priority system to determine which stimuli and events will enter our consciousness, based on issues of immediate safety and survival. Unconscious processing allows us to deal with more important tasks while leaving more constant and safe stimuli to be dealt with unconsciously.

We are generally less aware of the things that the brain does best, like vision and other sensory processes, and are most conscious of the functions it doesn't perform as easily—the more recent products of the TM, such as logic and mathematics. Still, even the Thinking Mind manages to carry out some advanced cognitive processes "offline." Studies show that we can unconsciously solve geometric problems without knowing how we do so. Over the past century, countless studies have demonstrated that visual characters such as letters or digits, geometric figures, and even the meaning of words can be perceived unconsciously.[1]

Consider how we learn any new skill—whether playing tennis, driving a new route, or typing on a computer. In each of these instances, we begin our education by being aware of every clumsy little detail, but over time, the processes become more automatic and unconscious.

The psychological term for these automatic processes carried on outside of our awareness is "habituation." If, while you were driving, you had to consciously attend to steering, braking, accelerating, judging distance and time, all the while having to watch motorists around you, you would quickly be overwhelmed—as many beginning drivers, in fact, are. The same is true of beginning musicians. A critical moment for anyone who plays music is moving from conscious to unconscious—in other words, from knowing it from the score to knowing it "in the fingers." No one could ever play music if she had to think about everything she was doing while moving from note to note.

It is somewhat ironic that, although we think of conscious awareness as the pinnacle of human evolution, much of our mental processing actually occurs unconsciously.

■ The Emotional Unconscious

While evidence for unconscious processing of cognitive information has been accepted for some time, scientific support for unconscious processing of emotions is much more recent. This is, unfortunately, typical of the many ways in which the Thinking Mind has long disparaged the Ancestral Mind. Modern psychology has usually been far more concerned with how we solve problems, dismissing the emotions as too "soft" a subject for serious research.

However, modern imaging techniques—particularly functional magnetic resonance imaging, or fMRI—have brought specificity, precision, and objective rigor to research on the emotions. In fMRI studies, people lie down and place their heads inside a powerful magnetic device that generates images of blood flowing through the brain. Blood carries glucose, and individual parts of the brain require extra energy in the form of glucose wherever they are activated. Whenever a structure is engaged in a particular function, the computer images created by the fMRI capture the increased blood flow to that area as splotches of light. This technology has enabled us to determine where different functions such as planning, problem-solving, and emotional responses actually take place in the brain.

Some of the earliest work on emotional processing dealt with subliminal perception, where research established that emotional stimuli that never registered in our awareness could still affect emotional behavior. In one typical study, emotional stimuli were presented by briefly flashing them on a screen with presentation times that were too short to allow conscious identification. Subjects nevertheless showed emotional responses in their autonomic nervous system, such as sweating palms. This could not have happened unless the emotional content of the subliminal stimuli had registered in the brain.[2]

Other subliminal perception studies demonstrated that we form emotional impressions of people without any conscious awareness of the basis for doing so. Such basic characteristics as facial expressions and physical features—whether skin color or voice intonation—are sufficient to activate emotions unconsciously.[3] For good or ill, then, a large number of the social judgments and inferences we make are unconscious processes.[4]

But it was the work of neuroscientist Joseph LeDoux, building on these earlier studies, that provided the physiological evidence for the key role played by unconscious processing in our emotional life. Central to the argument of this book, his research not only confirmed the functioning of the Ancestral Mind quite apart from the Thinking Mind, but also demonstrated that the brain itself gives priority to unconscious messages from this older part of us.

Working on fear responses in rats, LeDoux revealed that emotional stimuli that are relayed from the thalamus are sent simultaneously to both the amygdala, down in the Mammalian Brain (along the "low road," or thalamoamygdala pathway), and to the cerebral cortex, the seat of the higher cognitive functions (along the "high road," or thalamocortical pathway). The amygdala, being nonverbal, receives the input from the thalamus in the form of crude, almost archetypal information, a signal such as "predator" or "prey." But LeDoux's critical finding was that the input arrives at the amygdala *before* it reaches the cortex. What this means is that the amygdala allows us to intuitively "read" emotional stimuli and respond to them instantly, before we are consciously aware of precisely what we are responding to. This "quick and dirty" back road to the amygdala sacrifices accuracy for speed (reacting to "See stick ➤ Jump!," for example, when a more appropriate response would be, "See snake ➤ Jump!"), relying on the prefrontal cortex to refine the rough information and initiate a more detailed response at a more leisurely pace.[5] As LeDoux explains it, "You don't have to know exactly what something is to know that it is dangerous." In fact, some emotional stimuli may never even reach the neocortex or conscious awareness because the reticular formation does not always generate sufficient arousal to activate thalamocortical transmission or registration in working memory.

Extrapolating his finding into human terms, LeDoux demonstrated that natural selection placed its trust primarily in the unconscious processes of the Ancestral Mind, not the processes of the conscious Thinking Mind. It knew that in critical situations, the Ancestral Mind could be counted on as the wiser, more experienced observer and reliable guardian to keep us from harm.

In the emotional realm, Paul Whalen and his colleagues have shown that even photographs of facial expressions can unconsciously activate responses in the amygdala. The route between the thalamus and the amygdala may not be able to pass organized information such as the precise details of an entire face, but it does respond to the more basic facial features, such as the expression of the eyes.[6]

In any situation vital to the health and well-being of an organism, therefore, the intuitive reactions of the AM take neurological precedence over the slower, reflective reactions of the TM. Several points follow from this finding that are essential to the techniques of mind/body medicine, including those for accessing the Ancestral Mind:

1) Our emotional appraisals and reactions can occur unconsciously in the Ancestral Mind, yet still have a significant effect on our conscious feelings, moods, behavior, and physiology, *even when we have no idea that these processes are at work.*

2) We are not always aware of why we respond emotionally the way we do because emotional processes, including stress reactions, enter conscious awareness only in some instances. When those processes do reach the level of conscious awareness, they do so as thoughts, feelings, or moods through the prefrontal cortex and working memory, as the result of sufficient arousal generated by the reticular formation. Even then, however, we may not be aware of precisely why we feel the way we do, since the specific emotional stimulus that gives rise to the reaction may never reach conscious awareness in working memory. What this means is that, although we attribute our feelings to ongoing conscious experience (e.g., the current contents of working memory), they may actually occur for reasons of which we are unaware. In short, the source of our feelings may be very different from the rational reasons we employ to explain these feelings to ourselves. That's why we don't always consciously know why we feel anxious or have a headache or insomnia.

3) The fact that the amygdala reacts to and processes archetypal information, which is preverbal and symbolic, speaks to the potential power of imagery in communicating with the AM. Many of the therapeutic approaches to increased well-being we will offer are based on the fact that negative images can induce stress, while positive images can reduce it.

4) When someone engages in defensive body language, the language is the expression of an emotional response like anxiety or anger that is generated unconsciously. Just as we may be unaware of why we feel the way we do, we are also often unaware of the body signals that we send or receive, or the cause of the signals. But they register in the AM just the same.

If, for example, you snap at your spouse when you arrive home from work, you may rationalize that you did so because he was not listening to you. In fact, your angry outburst may have actually been the result of a series of stressors that you encountered during the day that were processed unconsciously. These unconscious stress responses that simmer below the level of awareness create a background readiness for fight or flight that heightens mental and physical arousal, effectively shortening your fuse and making your temper more likely to blow.

5) Such unconscious neurophysiological arousal, which can last for days, can be the basis for such conditions as:

• Heart disease. Stress has been linked to increased cholesterol, constriction of coronary arteries, myocardial ischemia, cardiac arrhythmias, ventricular fibrillation, and sudden cardiac death. Chronic hostility places men at greater risk for increased coronary blockage, heart attack and heart disease, and increased risk of dying from all causes. Hostile men are seven times more likely to die from any cause compared to less hostile men. In numerous laboratory studies, stress has been shown to increase blood pressure and heart rate, and reduce blood flow to the heart.

• Ulceration of the gastrointestinal tract, which can trigger symptoms of ulcerative colitis and inflammatory bowel disease. Stress alters gastric secretions and the contractions of the large and small intestine, and affects the time required for food to move through the gastrointestinal tract. It also alters bowel habits and can cause abdominal pain. Many of us notice changes in our gastrointestinal system under stressful conditions.

• Infectious diseases such as herpes, colds, and the flu. You may have realized that you are more prone to colds and illnesses when you are stressed; studies prove this is indeed the case. Stress compromises immune functioning, making us more susceptible to these illnesses. Studies show that married couples are more likely to come down with upper respiratory infections or colds in the days following intense marital conflict. Many other stressors, including job loss, marital separation and divorce, loneliness, academic exams, bereavement, and caring for family members with a debilitating illness such as Alzheimer's disease, have been shown to cause suppression of immune functioning. Although stress compromises immune functioning, it is unclear whether the magnitude of this effect is clinically significant (i.e., great enough to cause immunologic disorders).

• Panic disorder, post-traumatic stress disorder, major depression, and anorexia nervosa.

• Fibromyalgia and chronic fatigue syndrome.

• Tuberculosis, asthma, multiple sclerosis, arthritis, and diabetes.

• Inhibition of learning, memory, and problem-solving. Students who are stressed don't learn as well, for they do not process and handle new information as readily. (See Appendix B for a more detailed account of the effects of stress on health.)

Mind/body medicine has traditionally focused on conscious stresses, feelings, and thoughts to diagnose stress-related symptoms. Yet as we have just discovered, we often don't have conscious access to our emotional processing or the causes of our stressful feelings. Therefore, mind/body medicine has yet to address the true underlying brain mechanisms that can contribute either to illness or to well-being. As the last chapter has discussed, many of the techniques that will be presented here exert their therapeutic effects not just on the conscious TM but also on the unconscious AM by deactivating the neural circuitry of negative emotions and activating the circuits of health-enhancing positive emotions.

But before we turn to those techniques, we still need to dig a little deeper to increase our understanding of the Ancestral Mind's separate, distinctive world. In the next chapter, we'll learn more about the language that it speaks, the preverbal language of emotions, and the recently formulated scientific perspective on the power of positive emotions.

Chapter Four

The Power of Positive Emotions

Emotions, when they are uninflected by the Thinking Mind, are direct physical responses to the real world. They are always *about* something, whether a situation, an event, a person, or a thing.

When you become angry or excited, it may be the result of a signal from the Ancestral Mind warning you that something important is happening in the physical world that will have positive or negative implications for your well-being. But emotions can also occur in response to an image or a thought. Merely visualizing a traumatic event can elicit a strong emotional reaction.

No one definition of emotion has ever been universally accepted by psychologists, but there is a growing consensus that an emotion can be considered to be an "action tendency," or a *preparation to act* on the part of the Ancestral Mind concerning something that is significant to us.[1] Emotions motivate; they move us.

We don't typically choose to have emotions occur, but rather, experience them as happening *to* us. (If we could consciously control our emotions, we would not experience negative emotions very often.) Some emotions occur with such power that we feel ourselves "gripped" or "seized" by them. At the same time, we can place ourselves in situations, like socializing or buying a new car or going on a first date, that can influence the course of our emotions profoundly.

It's also true that emotion, in the form of intuition, plays a role in a number of areas thought of as purely rational. Many scientific discoveries, for example, emerged by way of hunches and sudden insights; some even came to scientists during dreams. Physicians often have to trust their instincts beyond any certainty science can provide when making decisions about a patient's diagnosis or treatment.[2]

As we've seen in the discussion of the "quick and dirty" pathway between thalamus and amygdala, evolution has placed a higher priority on immediate emotional responses than on thinking, reflection, and planning. In the wisdom of the Ancestral Mind, emotional reactions are more vital to our survival. Signals from the real world guided our behavior long before the Thinking Mind emerged. As Joseph LeDoux puts it, "Many emotions are the product of evolutionary wisdom, which probably has more intelligence than all human minds put together."[3]

In many circumstances we can't afford to let ourselves become bogged down with the prolonged deliberation of the TM, whose cognitive powers provide too many possible courses of action and repeated evaluations, which may be inconclusive and have dire consequences. If we always had to wait for the TM to determine whether something is dangerous, we might not only be wrong—we might be dead.

In his book *The Emotional Brain,* Joseph LeDoux asks, "What is irrational about responding to fearful situations with evolutionarily perfected reactions?"[4] Neurologist Antonio Damasio makes the same point in *Descartes' Error,* arguing that rapid emotional responses would have to be considered "rational" in emergency situations.[5] Yet, ironically, emotions are often labeled irrational simply because they do not involve the careful weighing of alternatives.

Although the AM's speed may result in outcomes that are not always perfect, the probability is still high that the time-tested, evolution-based responses are more adaptive in many situations than more inexperienced cognitive responses, which have a relatively brief evolutionary history.

■ Emotions, Feelings, and Moods

In our exploration of the physiology of the AM and Joseph LeDoux's groundbreaking research, we learned that emotional events initially capture our attention unconsciously (in the reticular formation, thalamus, and amygdala) so that we can begin to react physiologically to the relevant stimulus before we are consciously aware of it. If sufficient arousal is generated by the reticular formation, emotional responses (such as heart pounding and muscles tensing) enter conscious awareness in working memory as feelings and thoughts. Even then, however, we may not be aware of precisely why we feel anxious or irritable because, as we saw in subliminal perception research, the specific stimulus that caused the emotional response may never reach awareness through working memory. We may attribute our feelings to the ongoing contents of working memory when, in fact, our feelings may be the result of emotional stimuli of which we are unaware. (Recall the example of snapping at your spouse at the end of the day and rationalizing that your angry feelings were due to his not listening to you when, in fact, your angry feelings were due to stresses that were processed unconsciously earlier in the day.)

Emotions, then, are characterized by the automatic, physiological reactions that occur nonconsciously and rapidly (some emotions specialists believe that a true emotion lasts for only seconds[6]), and involve stress responses, facial expressions, and body language. Because the basic mechanisms of emotion were laid down long before consciousness appeared, they do not require consciousness for their expression.

Feelings, in contrast, are the *conscious* experience of emotional responses in working memory—the perceived outcome of unconscious emotional processing. Whereas emotions are defined by physiological responses, feelings are defined by their mental, experiential qualities. An example of the emotion of fright is my heart pounding, my muscles freezing, and my facial expression contorting if I encounter a large spider; my conscious awareness of my heart pounding and that I am afraid of spiders because they are dangerous is a conscious feeling of fear.

In a practical sense, emotions and feelings are so closely related that the terms are often used interchangeably. There is a third category, however, that we also need to understand—moods.

• Moods are often the *result* of emotions and feelings. Feelings of anger early in the morning can put us in a bad mood all day. Moods may in turn give rise to emotions like anger by making us tense, and thereby shortening our fuse.

• Whereas emotions are elicited by a particular stimulus, moods often lack an identifiable cause. We can often explain an emotion or feeling by pointing to a specific event, but when we are in a mood, we often cannot attribute it to anything specific.

• Moods are often more subtle than emotions (although moods can also be intense and overwhelm us), and we usually experience them as more diffuse and global. When we are in a foul mood, for example, *nothing* interests us. When we are in a good mood, *everything* looks rosy.

• While emotions are typically of short duration and intense, moods are often less forceful and last longer; they are a kind of background feeling that persists over time. If you get mad at your children, your facial reaction, stress response, and conscious feeling of anger typically don't last all day, but you may be in a bad mood for many hours. We don't spend a lot of our waking time experiencing emotions, but we do experience some kind of mood almost constantly.

• Conscious thoughts of the TM (such as "This is going to ruin my whole day") play an important role in generating moods and causing them to "brew" over time. (If I get into an argument, my initial emotional reaction is short-lived, but my internal monologue, which might consist of ongoing thoughts like "What a jerk" and "I can't believe he did that to me," can lead to a bad mood.) This is where the TM's negative internal monologue plays a strong role.

Just as emotions have taken a backseat to cognition and problem-solving as subjects of scientific research, so, too, have researchers been imbalanced in their approach to emotions per se. While over thirteen thousand papers on negative emotions are cited in the psychological literature, those on positive emotions scarcely exist.[7]

Positive emotions are associated with things that are inherently pleasing to the senses or that represent an opportunity or the achievement of a goal. Negative emotions involve things we withdraw from because they may be harmful, threatening, or prevent us from achieving a goal.[8] Both these cues have their value in an evolutionary context. Positive emotions tell us to continue to do what we are doing; negative emotions alert us to take steps to "set things right." This is the foundation for the simple wisdom of the Ancestral Mind. The trick is in clearing away the distraction of the Thinking Mind to attend to the Ancestral Mind's messages.

Most scientists agree that there are at least four basic emotions: joy, anger, sadness, and fear, of which there are many variations.[9] Some of the variations of joy, for example, include happiness, ecstasy, satisfaction, rapture, thrill, pride, delight, contentment, and bliss. Many complex emotions are uniquely human, for they require the participation of the Thinking Mind. Shame or embarrassment, for example, are negative emotions that reflect psychological injuries to one's sense of self—the provenance of the Thinking Mind.

Research by psychologist Richard Davidson, a leading expert on emotions at the University of Wisconsin, suggests that greater activation of the left prefrontal cortex heightens positive emotions, while increased activation in the right PFC heightens negative emotions.[10] If you are an optimist, you are more likely to exhibit greater activity in the left PFC; a pessimist will tend to exhibit greater right PFC activation. Individuals who have too little left PFC activation or too much right PFC activation cannot "turn off" negative emotions or activate positive emotions as readily and are therefore less able to handle stress. In several studies, Davidson has also documented that people with a history of clinical depression display more activity in the right PFC and less activity in the left PFC.[11] Although Davidson's findings are based primarily on female subjects, I have extended his work and demonstrated that the relationship between PFC activation and emotional style can also be measured in men.[12]

Each basic emotion has served a primary function throughout our evolutionary history. Sadness helps us to disengage from life and conserve energy so that we can adjust to a loss (such as a death) or cope with a disappointment. With the drop in interest in life's pleasures and activities, we are better able to ponder the meaning of loss, adapt to its consequences, and plan accordingly. Anger energizes us to protect ourselves and those close to us, right a wrong, or persevere when our goals have been blocked. Fear creates a highly adaptive vigilance for potential threats; it also promotes escape by triggering the stress response that has been essential for survival throughout our evolution. Even fear's close relative, worry, which is distinguished from fear by the lack of a real threat as a motivation (we can worry about things that don't exist), can sometimes serve as a helpful rehearsal for what might go wrong and how to deal with it. Joy signals that we have achieved a goal and, like many positive emotions, promotes prosocial behaviors like altruism.

Negative emotions in response to signals from the outside world are a normal and necessary part of life. They often keep us alive by keeping us alert and can motivate us to make important life changes. Persistent negative emotions generated by signals from the Thinking Mind are destructive, however, and must be kept in balance with positive emotions. It is in its capacity for generating these positives that the Ancestral Mind can play a vital role.

Psychologists Ed Diener and Randy Larsen have found that the simple ratio of pleasant to unpleasant emotions is a central factor in how people evaluate their lives. Diener's laboratory has determined that the degree of basic emotional pleasantness in daily life is a stronger predictor of life satisfaction than physical pleasure, satisfaction within specific domains, or goal achievement.[13] It seems that no matter how many classic romantic scenarios suggest otherwise, intense emotions such as ecstasy are not nearly as important as the day-to-day experience of emotions that are even mildly positive.

When people are able to keep distressing emotions in check and counterbalance them with positive ones, we call this a sense of well-being. From an evolutionary perspective, we know that no negative emotion, such as fear, could have been adaptive for more than an acute period of time. A perpetual state of fear would have interfered with the business of survival, by actively preventing such activities as altruism, exploration, food collection, and mating.

Although occasional worry can be helpful in our lives, chronic worry is a comparably maladaptive negative emotion—in this case, one generated by the Thinking Mind. It disrupts our connection to the Ancestral Mind and thus to the normal continuum that helps achieve a balance between positive and negative emotions. When we are anxious, we can't concentrate and are unable to experience positive emotions. Anxiety can even become pathological when it impairs our motivation and ability to perform daily activities. Like anxiety, anger can become harmful by turning into hostility and creating a chronic state of tension that disrupts relationships, enjoyment of life, and health.

Positive emotions do more than draw us toward things that are pleasant, for they are ultimately fundamental to existence. Positive emotions communicate to the mind and body in a life-enhancing way, energize us, and enrich our lives. We have an innate desire to experience them, as they improve health and well-being by:

• Deactivating structures in the AM in order to produce a state of physiological quiescence, and, in some cases, serenity and stillness. This state, as we will see, restores energy, quiets the TM, and counters the physiological arousal of the stress response and its many deleterious health consequences. In contrast, a number of positive emotions like happiness can involve a stimulus that captures our attention and *activates* structures in the AM (ecstasy, for example, can be a highly excitatory emotional state); these activating positive emotions likely exert their therapeutic effects by turning off the circuitry of negative emotions and the stress response, and by allowing us to recover from stress.[14]

• Inhibiting negative emotions, enhancing coping, and improving our outlook on life. Positive emotions can help us to face obstacles with equanimity.

• Increasing energy for achieving goals, fostering a sense of exploration and discovery, and promoting greater enthusiasm for life's tasks.

• Eliciting feelings of something greater than one's self, including feelings of a supreme power, by altering the contents of working memory and the orientation cortex that maintains the self's boundaries.

• Enhancing prosocial behaviors like openness, bonding, cooperation, sharing, altruism, and love and a readiness for peaceful interaction.

• Enhancing clarity to improve mental functioning, enhancing creativity and problem-solving abilities, and improving performance and productivity.

Experiencing just a few moments of positive emotions can carry us through the day by absorbing us and distracting us from the troubles of life. Emotions like reverence or awe are so powerful that they can become landmark memories of what life can be like. Much of the effort we expend in accessing the Ancestral Mind is to benefit from its ability to induce the positive emotions that can enhance our mental and physical well-being.

■ Nature, Nurture, and Emotions

We all know people who are characteristically optimistic and a pleasure to be around. Then there are the complainers who tend to "awfulize" about their exhaustion, how much they have to do and how little time they have to do it, and how tough things are generally. They are alarmists who see catastrophes in everything and feel a need to share their "doom-and-gloom" outlook on life with everyone.

Why are some people more positive in their emotional style while others are more negative? Genetics plays a role. Research on twins demonstrates that a significant part of personality is inherited. Additionally, we have already seen that individuals differ in the relative activity of the left and right frontal lobes of the brain; this difference correlates with emotional style and can be measured in infants.[15] Richard Davidson and Nathan Fox found that individual differences in frontal brain asymmetry are present within the first year of life and predict important aspects of an infant's response to stressful challenges such as brief episodes of maternal separation. In a similar vein, Harvard developmental psychologist Jerome Kagan has demonstrated what many of us have observed in our own families—that infants exhibit distinct temperaments (such as timid or bold), which are correlated with distinct patterns of brain activity.[16]

Each individual's Ancestral Mind is not a generic, unvarying, one-size-fits-all model. Not only are we each endowed with a tendency toward positive or negative emotional styles, but the environment in which we live, particularly in childhood, is also part of the equation. A child born with a tendency toward a negative emotional style but raised in a nurturing environment in which she wasn't exposed to highly stressful events might never exhibit that negative emotional style. Likewise, children who are exposed to stressful or traumatic experiences early in life are at greater risk for emotional disorders. (A childhood history of abuse, neglect, or trauma even increases later susceptibility to gastrointestinal disorders.)[17] In short, while genetics may predispose us to a specific emotional style, in the absence of a certain level of environmental stress, that predisposition may never express itself.[18]

Regardless of genetic influences and the world in which we grow up and live, by learning to access the Ancestral Mind we can tip the emotional scale toward the positive. Daniel Goleman's book *Emotional Intelligence* suggests that, with the correct training in emotional skills like self-awareness, impulse control, and empathy, temperament is not destiny: emotional styles can be changed for the better in children. For example, studies on infants reveal that a good percentage of those with timid temperaments lose their timidity by five years of age.[19] Happily, research on psychotherapy, including cognitive behavioral therapies that incorporate relaxation and cognitive restructuring techniques, has also provided convincing evidence that emotional styles can be changed in adults.[20]

We owe a debt to Freud for the fundamental insights that led to modern psychotherapy, but as Dr. Seymour Epstein, a psychologist at the University of Massachusetts, points out, the critical weakness in Freud's underlying theory of an emotionally tormented unconscious is that it makes little sense from an evolutionary perspective.[21] In writing about emotion as if it consisted only of dangerous, unbridled passion, Freud failed to consider that the emotional unconscious developed an intuitive wisdom of its own as a result of overseeing survival for so long in the course of human history. The fact that Freud's system of psychoanalysis was developed during his studies of seriously disturbed people also contributed to his view of the pathological nature of emotions, as did the long Western tradition of viewing man as essentially competitive and isolated and needing to control his baser impulses by reason.

But far from being a morass of repressed sexual desires and vices, the emotional unconscious is an essential, adaptive part of the human mind. As we have seen, during the long evolution of the Ancestral Mind, emotions have served as our internal guides to daily living. They motivate us to act, warning us that something important is happening, and that we should take whatever steps are necessary because our well-being is at stake. Emotions draw us toward certain people, things, behaviors, and ideas and push us away from others. There are many important life decisions that cannot be made solely on the basis of rationality; emotions and intuitive, gut reactions also supply us with the wisdom to navigate our most important decisions, including whom to trust, whom we should marry, and what work we should pursue. In a very real sense, we think by feeling.

In the next chapter, we're going to examine in greater detail how ancestral man was shaped by a balance of positive and negative emotions, and how the Ancestral Mind, operating on the basis of sensory and emotional information, and firmly grounded in the real world, kept him in a state of healthful equilibrium.

It was the Ancestral Mind, after all, that enabled our forebears to overcome intimidating physical challenges and develop a lifestyle that was so highly successful that humans emerged as the dominant species on earth. Although evolutionary success is usually attributed to the development of language and tools, emotions also played a crucial role. They served as survival and communication mechanisms that protected us from harm; facilitated sharing, cooperation, and nonverbal communication; attracted us to food and mates; and, pointed us toward things that enhanced survival. We owe much of our existence as a species to the wisdom of the AM and emotions.

For most of our evolution, the activation of negative emotions in the AM (such as fear) were triggered by stressors that were acute, concrete, physical—and short in duration. The TM-based consciousness that allows for the existence of such things as anxiety, a negative internal monologue, and stresses that are chronic, diffuse, and abstract, which we now take for granted as "normal," is, in reality, the product of a very recent period of our development, and the blinking of an eye in terms of evolutionary time. The fact that evolutionary change occurs slowly, across thousands of generations, means that it is unlikely that the AM will adapt to the novel aspects of modern life— the TM's stresses—anytime soon. We are, and will long be, designed for the life of our ancestors and the AM, not the modern world of the TM. In other words, the modern world of the TM interferes with the evolutionary design of the mind. *The root cause of modern stress is the discrepancy between TM consciousness and AM consciousness, between modern world and ancestral world.*

The trouble is, the TM, with its incessant internal monologue, great cognitive powers, and strong sense of self, has seduced us into believing it is not just part of us but *is* us. Reinforced by a culture that equates happiness with advances in technology and materialism, and which views the TM as the key to our comfort, well-being, and betterment of life, the TM now forms the basis of our most fundamental beliefs about the world and has become our sole arbitrator of reality. Even the Latin term for our species, *sapiens,* "the wise one," reflects the emphasis on cognitive ability in how we view ourselves. We have forgotten that another mind, intelligence, and consciousness exists.

Now, no one would propose that we return to a hunter-gatherer lifestyle as a solution to the stress of modernity. The point, rather, is to reanimate ancestral states of mind and the positive aspects of evolutionary heritage so that a better "fit" exists between modern life and ancestral life, between our mental world and that of our ancestors. By minimizing the discrepancy between the world we live in and the world we evolved to live in we can restore at least some of the equilibrium that our well-being demands.

It was only with the advent of the self-conscious Thinking Mind, its free will and its technological sophistication, that we were able to step beyond the direct constraints of natural selection and manage to develop such maladaptive traits as persistent anxiety. Self-awareness and advanced cognitive abilities have given us a great many advantages—advantages that promote survival of larger numbers of inhabitants on this planet than would be possible with the simple but efficient methods of ancient man. But that progress has come at a great cost, for by creating maladaptive negative emotions and inhibiting positive ones, we have disrupted nature's balance. The result is a worldwide increase in depression, a collective emotional crisis, and an existential malaise. Even our children are affected, for they are more lonely, troubled, and depressed than previous generations.[22]

Most of the mind/body techniques developed to help us cope exert their therapeutic effects either by directly affecting the body (by minimizing the stress response) or by altering conscious feelings, thoughts, and behaviors. Here, we will add other approaches that work in a more fundamental way, techniques that offset the unnatural impulses we have had to adopt to keep up with the modern world, techniques that enable us to consciously reanimate the AM and ancestral states of mind, and to reconnect with the most positive aspects of ancestral life.

Chapter Five

The Ancestral Way of Life

How did our ancient ancestors manage not only to endure but to prevail against famine, floods, and other calamities of everyday existence at a time when they had so little to protect them against the elements?

They did so in large part by relying on the emotions, behaviors, and attitudes that were shaped by natural selection. We've seen that emotions, unlike thoughts, are sensory based and correlated with immediate reality. Our emotional responses developed over millions of years precisely as a messaging system to keep us in synch with our outside environment. They guide us from moment to moment, and when they are unfettered by the contortions of the TM, which can obscure the present through the relentlessness of its focus on the future or the past, they give us a more reliable picture of our surroundings.

We've also seen that many of the behaviors based in the AM, such as non-verbal communication, were crucial to survival. And we will see that the AM's adaptive attitudes facilitated survival by maintaining a balance of optimism and (at times) realistic pessimism.

If these ingrained emotions, behaviors, and attitudes, wired into the neural circuitry of the Ancestral Mind, had not made our ancient ancestors stress-hardy by promoting health and well being, they (and we) never would have survived the long evolutionary trek in which traits that enhance fitness are passed along generation after generation, and traits that make us less fit die out (along with the creatures that carry those negative traits).

As we prepare to learn to reconnect with the Ancestral Mind, it will help us to have a clearer picture of precisely the kind of life the AM presided over before the TM came along. This is an especially important consideration, given that the Thinking Mind, with its obsession with material "progress," has done so thorough a job of denigrating and devaluing its older partner.

It was during the height of the Age of Reason, shortly after Descartes proposed, "I think, therefore I am," that the English philosopher Thomas Hobbes described the ancestral human condition as "solitary, poor, nasty, brutish, and short." In his view, the primordial way of life was little more than "the war of every man against every man."

Armed with germs and steel and Hobbes's mindset, readily dismissing any and all "less advanced" peoples because of their less sophisticated technology, Europeans went on a centuries-long binge, destroying indigenous cultures without regard to their value or the wisdom they contained. The only distinction they saw was "primitive" versus "civilized." The Thinking Mind completely overlooked the fact that these "primitive" peoples, guided much more by the Ancestral Mind, might have understood the world more deeply and inhabited it in a far more adaptive way.

While there's no doubt that ancestral life contained daunting physical hardships, according to the most recent interpretations of the anthropological evidence, the world before the Thinking Mind emerged was characterized not just by stress-hardiness and social support, but by a kind of cooperative attitude that could best be described as altruism. Attribution of such qualities to our ancestors is not based on some idealized notion of the "noble savage," but on the simple fact that those characteristics were more adaptive, meaning that they worked better than other ways of being to ensure survival. Enhanced survival meant greater reproductive success, which meant more offspring carrying those same traits.

From what we have learned about the Ancestral Mind, we can also infer that ancestral life was vibrant and intense in ways that we might well envy today. Unbiased by internal monologue and rational self-consciousness, ancestral man's awareness was profoundly merged with the present. Because he did not see himself as a distinct entity separate from the rest of creation, he would not have felt isolated and existentially lonely. We can infer that he experienced an integral connection with the environment, in which animate objects were perceived as alive and personal, and in which mystery and divinity were perceived in ordinary phenomena. Closely tied to a small band of related people, and to a particular place, his life had a clear sense of purpose—the survival of himself and his family—which he achieved by responding appropriately to environmental cues.

Negative impressions of anything associated with ancestral life have continued to persist, however, in part because the Thinking Mind simply had its facts wrong.

The view of primitive man that assumes a dark, aggressive, and violent nature was propagated as recently as the 1960s by an Australian anatomist named Raymond Dart. In a series of papers, he reviewed anthropological evidence and concluded that our ancestors were "carnivorous creatures that seized living quarries by violence, battered them to death, tore apart their broken bodies, dismembered them from limb to limb, slaking their ravenous thirst with the hot blood of victims and greedily devouring livid writhing flesh."[1]

But that brutal portrait of "the killer ape" simply has not held up. According to world-renowned anthropologist Richard Leakey, our ancestors actually depended far more on gathering than on hunting for their sustenance.[2] And because the Ancestral Mind was so finely attuned to the physical environment, labor was extremely efficient. Foragers could collect sufficient food for the day in three or four hours, which, in fact, gave them a good deal more leisure than we enjoy today.[3] Most significant, the requirements of successful gathering demanded a social system that, unlike our own individualistic society, fostered cooperation and interdependence. In a major article published in Scientific American in the late 1970s, Harvard archaeologist Glynn Isaac helped shift the emphasis away from hunting as the force that shaped human nature.[4] In Isaac's view, it was the regular sharing of food between mother and offspring, and the systematic expansion of the sharing network to include adult females giving to adult males, that enabled man to survive in a harsh and indifferent world. Isaac cited archaeological evidence that suggests that, although meat was an important component of the ancestral diet, it was more likely acquired by scavenging than by hunting. Scavenging, too, could be a very efficient activity—an abandoned leopard kill could supply a day's nutrition and require only half an hour's work.

A conference in 1966 at the University of Chicago on "Man the Hunter" had already acknowledged that the gathering of plant foods provided a significant source of calories for most hunter-gatherer societies. Tools, one of the major technological breakthroughs that would later make hunting more efficient, were first used for gathering plants, eggs, honey, termites, and ants and probably for catching small burrowing animals. The first tools associated with animal prey were very likely those that allowed humans to smash the bones of scavenged carcasses in order to expose the fat-rich marrow and to cut meat into edible pieces.[5]

Of course, asserting that sharing and cooperation were fundamental elements of human evolution is not to claim that ancestral man always lived up to some romanticized utopian ideal.[6] What it does mean is that the image of relentless brutality simply doesn't fit the evidence. According to Richard Leakey, the "killer ape" paradigm "is one of the most dangerous and destructive ideas that mankind has ever had."

In fact, ancestral humans were probably no more disposed to aggressive behavior than any other species. According to L. H. Kelley, author of *War Before Civilization* (1996), violence in early societies was the exception rather than the norm.[7] Most prehistoric warfare was a response to unpredictable disasters that corresponded with ecological and climatic changes, putting such intense pressures on bands of humans that they were forced to take from others in order to survive. The majority of the time, peacefulness was simply the much better option.

Anthropologists point out that it was only when organized groups began to build cities and demarcate precise boundaries for ownership of farmland that we find more consistent evidence of warfare and intentionally inflicted death. It is no coincidence that this turn toward organized violence began around the same time as the emergence of an individualized self-consciousness capable of detached, abstract thinking.

The broader point to be made here is that from an evolutionary perspective, hyperaggressiveness is no more adaptive than persistent hyperanxiety. Survival in the wild undoubtedly required a certain degree of assertiveness and fortitude, but, again, according to the evolutionary record, long-term survival depended far more on cooperation, as well as a fine-tuned harmony with the environment, than it did on hostility and on claiming "dominion."

Ironically, the pop culture idea of the brutish "caveman" is based largely on false assumptions about the Neanderthal and Cro-Magnon people of Ice Age Europe.[8] Neanderthals thrived in the Neander Valley of Germany 75,000 years ago; Cro-Magnons, named for the locality in France where they were discovered, lived during a span from about 50,000 years ago to perhaps 10,000 years ago.

Far from being insensitive brutes, though, these ancient protohumans, guided by aspects of the Ancestral Mind, presided over the birth of Western art and music. Given that fragments of flutes have been found at Neanderthal sites, and given that language is thought to have developed as recently as 35,000 years ago, it is possible that Neanderthal man played instrumental music before he spoke.[9]

Cro-Magnon cave painting, offering even richer insights into the workings of the Ancestral Mind, reached its greatest flowering on the stone walls of the Lascaux cave in France some 15,000 years ago. Vivid depictions of horses, stags, bulls, as well as the human form, were painted on crystalline surfaces in bold images filled with life. Exquisite carved statuettes of horses have been recovered from sites known to be 32,000 years old. Small ivory beads and human figures have been dated to around 30,000 years ago.

Some of these products of the Ancestral Mind contain recurring motifs that still have currency in art and literature today. Many thinkers interpret this continuity as reflecting unconscious emotional memories wired into the brain as nonverbal, sensory images. These repeating mental patterns, called archetypes, are the basis for the school of psychology founded in the twentieth century by Carl Jung. Like the song wired into the brain of a young bird, certain images—the father figure or "king," the mother figure or "queen," the lover, the figure of the trickster who challenges the established order—have found expression in almost all cultures. They are the common threads in each society's otherwise distinctive myths and legends.

While no people living today can give us an exact picture of what life was like before the advent of the Thinking Mind, a major research project conducted in the 1960s and 1970s offered a rare opportunity to observe a hunter-gatherer society still intact, when a team of Harvard anthropologists studied the !Kung San tribe of Botswana.[10] Firsthand observation of these people seemingly lost in time has confirmed the findings of earlier archaeological research based on artifacts from the distant past.

Despite their extremely marginal, desert environment, the !Kung made a reasonable living late into the twentieth century. The social unit consisted of small, mobile bands of about two dozen individuals—males, females, and their offspring. These bands interacted with others, forming a larger social network. Skilled foragers, the twentieth-century !Kung were still closely attuned to the environment, reading it more closely than we could ever comprehend. The males scavenged or hunted, and the females gathered plant foods. Because the availability of meat was always so uncertain, plants provided the largest proportion of their diet.

In *Nisa: The Diary of a !Kung Woman*, anthropologist Marjorie Shostak describes how the !Kung stressed equality and discouraged self-importance. Their emotional fulfillment came through relations with one another rather than through their possessions. Their culture was carried in their heads, not on their backs. They had a large amount of spare time which was spent visiting and entertaining and sharing food. The !Kung were great storytellers, and they loved to laugh. According to anthropologists, the hunter-gatherer existence of the !Kung can be considered the original affluent lifestyle, in which all the people's needs were easily satisfied.[11]

Let me stress once again that I'm not proposing that the secret to happiness lies in living in the wild and foraging for food. Instead, my point is that we should intelligently enjoy what we can from the Thinking Mind's material "progress" while at the same time benefiting from the Ancestral Mind's instinctive ways of promoting physical and emotional well-being. We don't have to return to the caves of Europe or the plains of Africa to access the parts of our brains that promote that groundedness and well-being.

While we can derive great benefit from a return to certain aspects of the ancestral life—periods of exercise and relaxation, greater exposure to sunlight and wilderness, and both solitude and social support—it is primarily the ancestral *state of mind* that we need to learn to access. This is the mental state in which we learn to bypass the TM's linguistic screen and the subject/object relationship that normally filters and restricts our perceptions. By doing so we can learn to participate in experience, as our distant ancestors did, rather than merely observe it. By doing so we can lose our everyday sense of self and gain a much more immediate and intense grasp of our own existence as an integral part of a much greater reality.[12]

In subsequent chapters we will examine more closely certain human characteristics that developed out of the way of life our ancestors led for millions of years, characteristics such as a deeply rooted preference for rich social networks, for laughter, for optimism, and for positive illusions, as well as a limited tolerance for anger and pessimism. But even the ancestral state of mind itself, simply *as* a state of mind, is immensely beneficial.

Accessing the Ancestral Mind helps us to be clear and focused. It makes our senses and perceptions keener and richer. It offers a state of being rather than thinking, in which awareness merges with the present. It is a state of absorbed attention, in which a person is completely there, totally immersed, like the child who spends not minutes or hours but an entire day playing by himself in the sand at the beach, pouring, sifting, digging, seeing and creating patterns, building castles.

A sojourn into the ancestral state of mind for just a few moments daily can change our perception for the rest of the day. These respites can allow us to temporarily suspend TM consciousness, then bring the benefits of the AM back with us into the ordinary moments. They energize us and can help us to perceive ourselves and our lives in a new, richer way.

In recent years psychologists have explored such transcendent states of mind under various rubrics. After interviewing thousands of individuals over the past twenty years, Mihaly Csikszentmihalyi has described a characteristic mental state that he calls "flow." In flow, an individual is so absorbed in the moment that excellence streams effortlessly, awareness merges with actions, and the individual feels immersed in a state of relaxation, with access to maximal energy. The mind does not wander, and is undistracted by awareness of the past or the future or by any unpleasant aspects of life.

In *Flow: The Psychology of Optimal Experience,* Csikszentmihalyi describes one of the typical characteristics of flow as the sense that time no longer seems to pass the way it usually does. The objective, external duration we measure with the progression of the clock is rendered irrelevant in flow. When they have a flow experience, most people report that time proceeds much faster than usual; hours seem to pass by in minutes. But it's not just the boundaries of time that seem to dissolve. In Csikszentmihalyi's words:

> What slips below the threshold of awareness is the concept of self. People stop being aware of themselves as separate from the actions they are performing. And being able to forget temporarily who we are seems to be very enjoyable. Loss of the sense of a self separate from the world around it can lead to a feeling of union with the environment, self-transcendence, and a feeling that the boundaries of our being have been pushed forward, so that the person is transported into a new reality, to previously undreamed-of states of consciousness. The hallmark feelings that result from this state of consciousness are joy, awe and wonderment.[13]

Csikszentmihaly uses terms like "grace" and "ecstasy" to describe this experience in people from all walks of life. He points out that, although flow occurs most often when people are engaging in a favored activity like rock climbing or composing, it can also take place spontaneously in response to something as simple as listening to a piece of music or witnessing something beautiful in nature. He notes that all of us experience flow from time to time and that among the keys to achieving flow are the focusing of attention and relaxation, which are the very same prerequisites to entering meditative states, as described in Eastern traditions. Although we lack firsthand accounts from the !Kung or any other primitive people that describe the kind of life they live or lived in terms of "flow," making that connection is not a great leap.

Naturalist Bernd Heinrich, writing in *A Year in the Maine Woods*, described sitting during a hunt in a deer blind, watching and waiting. He was surrounded by nature and eventually became absorbed in it. And then, for an indeterminate amount of time, he *was* nature. He realized after the fact that more than an hour had passed without his having had a single rational thought. The poet Andrew Marvell speaks similarly of "a green thought in a green shade." Ancestral Man, depending entirely on his Ancestral Mind in the ancestral state, had it even better: There was no self-reflective, detached, and abstracted mode of thought at all.

The late humanistic psychologist Abraham Maslow spent the latter part of his career studying and writing extensively about another dimension of the ancestral state of mind, which he called the "peak experience." When Maslow broke new ground by beginning to explore the psychology of health, he found that emotionally healthy individuals, those whom he called "self-actualizing," tended to report having moments of great awe, rapture, or bliss. During these peak experiences, the individual feels more unified and whole, at the peak of his powers. Peak experiences represented the individual's happiest, healthiest, most fulfilling moments in which, once again, self-consciousness and all separateness from the world at large dissolved.

Compare what we've observed about the ancestral state of mind with how Maslow describes characteristics of the peak experience:[14]

1. Perception is relatively ego-less. The individual fuses with objects into a new, larger whole. Objects are seen as free from relations, purpose, or usefulness to anything else.

2. Awareness of the past and future is lost. The person lives only in the moment, totally immersed in the here-and-now. A distortion of time and space occurs.

3. Normal, everyday consciousness is widened and enriched, and one feels that one has experienced a "higher," more direct state of consciousness, that one has perceived the true essence of things. This altered state of consciousness has a lasting effect on the individual, for it can be summoned in the duller moments of ordinary life.

Maslow found that "one can and does learn from these experiences that joy, ecstasy, and rapture do in fact exist and that they are available to the individual." He also concluded that being overly rational or materialistic—in other words, living too much in the Thinking Mind—prevents one from having peak experiences. Far from sharing the modern prejudice against the ancestral and the primitive, he characterizes the inability to have peak experiences as a lower, lesser state in which we are not fully human.

During the earlier phase of Maslow's study he remained skeptical about "mystical" experiences, dismissing them along with accounts of the supernatural, until he realized that emotionally healthy individuals were reporting such experiences consistently. Having once "sniffed at them in disbelief and considered them nonsense, maybe hallucinations—almost pathological," he now regarded peak experiences as "moments of pure, positive happiness, when all doubts, all weaknesses were left behind."

Flow, peak experiences, and what others call mystical experiences are all qualitatively different from our normal TM-based self-consciousness. Each of them is characterized by a sense of unity or oneness, the feeling of the personal self becoming less real (or even disappearing), a sense of timelessness and a sense that the horizon of awareness has been greatly expanded, and feelings of reverence, awe, and ineffability.[15]

As we discussed in chapter two, Andrew Newberg's work on the orientation association cortex provides a physiological explanation for the mystic's sense of union with the universe, the ultimate feeling of transcendence often achieved through meditation, chanting, or even ecstatic dancing, all of which involve introducing repetitive sensory input into the orientation association cortex. When the orientation association cortex receives sufficient

stimulation of this type, it simply shuts down. As a result, the brain's perception of bodily boundaries and the self dissolves.

A hallmark of the mystical state is an intuitive sense of "realness" in
which one "sees" rather than thinks. The world is perceived more directly,
and the vividness and richness of normal waking consciousness are greatly
enhanced. This experience is described by mystics and theologians as the
highest state of consciousness, in which the usual sense of self is swept away
by a new, vaster perception.[16]

These initial insights regarding flow and peak experiences have been
subsequently pursued by emotions researchers. Some scientists believe that
mystical awe may constitute a basic emotion with its own physiology: a sensation of expansion in the chest, chills, welling up of the tear ducts, and
clenching of the throat. Awe may function as a kind of adaptive moral inspiration that motivates people to be more caring and giving. In this sense,
morality may be an inherent part of the Ancestral Mind, driven by instinct
and emotion.[17]

Emotions researcher Jaak Panksepp believes that the intensely caring and
generous attitude called altruism played an integral role in the development
and expansion of the human brain.[18] As we have seen, our ancestors survived
not by ripping one another apart but by helping one another hunt, gather
food, and defend against predators. The simple act of one individual's keeping watch while others sleep has huge survival advantages. Altruism facilitated all forms of cooperation, sharing, and communication—key forces that
shaped human evolution. Psychologist Jonathan Haight uses the term "elevation" to describe this moral inspiration that motivates people to be more
altruistic. Because of its distinct physiology and motivational effects, he believes elevation should also be considered a distinct emotion.[19]

The fact that altruism is not unique to humans only strengthens the argument that it is wired into us, an instinct that survived the process of natural
selection because it served an adaptive function. Chimpanzees, for example,
console, share, mediate, and reconcile after conflict. Strong arguments have
been made that these prosocial behaviors served as the foundation for human morality according to the theory of "reciprocal altruism," in which by
trading favors, closely related individuals ensure the survival of at least some
of the genes that they have in common.[20]

Other researchers believe that elevation may be a kind of awe, which can inspire a feeling of being in the presence of something greater than oneself. Great religious leaders may have had such a powerful effect on people, in part, because the leaders inspired feelings of awe or elevation.[21]

The tragedy of our technologically advanced civilization is that, while it protects us from the harshest negatives of ancestral life, it also cuts us off from the most inspiring positives. Peak experiences are few and far between when we find ourselves commuting along the same highway or subway route every day. The exhilaration of flow in one's daily work is limited to a lucky few. And we rarely encounter awe or ecstasy relaxing in our suburban living rooms, no matter how many channels of cable we have. Instead of brilliant sunsets and a sky full of stars we have *The Sopranos*. Instead of a calming silence we have talk radio. Instead of the rhythms of the natural world we have instant messaging. And rather than surrendering to reverie and relaxation and the joy of simply being alive, we feel guilty whenever we aren't "accomplishing" something.

The ancestral state of mind is evidence that we live in a world of greater mystery than our Western worldview has allowed us to experience. It can enable us to reexperience a sense of joy, wonder, enchantment, and a world with which we have lost touch. In this state of consciousness, the mundane world of worries and concerns can temporarily disappear into the background and give way to another dimension of fulfilling, transcendent experience: beauty, mystery, even rapture. These awe-inspiring respites can stretch the limits of belief and nourish the mind forever by becoming landmark memories of what life can be like below the surface of awareness. They promote a very old and deep assurance that life has continuity and meaning, that things are somehow in place.

There's no question that the life of early humans tens of thousands of years ago carried with it a level of physical hardship that few of us would willingly return to. And yet we have paid a high price for our comforts.

Our lives are no longer measured by the natural increments of the sun and moon and seasons and their ritual observance, but by the industrial time clock. Passage through life is measured far less in relation to family, culture, and tribe than in relation to preparation for work, working, and retiring from work. We are units of production first, and human beings a distant second. And driving it all is the TM's incessant monologue, whose pulsing message throughout adult life is, essentially, "Worry and work, worry and work."

In the chapters that follow, we are going to explore how to restore communication with the Ancestral Mind and draw upon its deep store of emotional wisdom and profoundly stabilizing, recurrent images. We're going to heed the medical facts that deep relaxation and rest revitalize us, solitude restores and nature inspires, physical activity and bright light energize, life involves rhythm and being absorbed in the here and now, and positive emotions are essential to health.

PART II

Reclaiming the Power
of the Ancestral Mind

Chapter Six

Taming Toxic Thoughts

Kurt had worked for Allied Industrial for three years. He liked his job and his employer liked him—his most recent performance review was filled with goals "Achieved" and even an "Exceeded" or two. But times were tough, the company was downsizing, and he was told on only two weeks' notice that his position was being eliminated. Suddenly, Kurt's bright future looked to him like the end of the world.

"What'd I do to deserve this?" Kurt's TM started in.

"My luck always stinks."

"I'll never find another job like this one, not in this economy."

Kurt's negative outlook led to a month of anxiety and depression. He couldn't concentrate or find the motivation to overcome his sense of doom. He slept poorly and developed headaches, but at last he put his résumé together and started looking for a job. To his surprise, he quickly landed several interviews and received two offers. The position he ultimately accepted included a significant pay increase over his previous job, a move to a part of the country that he loved, and a company car. Six months later, Kurt was promoted to manager, a step up that brought with it a very nice raise.

To a significant extent, we are each responsible for the creation of our own experience of the world. Despite its reputation for logic and linear analysis, the

Thinking Mind all too easily spins out of control, concocting scenarios that exaggerate the negative and fail to account for the positive.

But as we've seen, there is deep down within us, in the Ancestral Mind, a more grounded Self with a powerful instinct for well-being—an instinct that can help us overcome the TM's imbalances. That Self is the source of a will with which we can exert control over our own thoughts and redirect them toward a healthier worldview. Although we can't always eliminate the stresses in our lives, with the proper effort, we can at least modify our response to them.

A primary technique for exerting control over our internal monologue and quieting the Thinking Mind is called *cognitive restructuring*. This method is based on research that shows that negative thoughts create negative emotions, which activate the neural circuitry of stress. That stress, in turn, leads to even more negative emotions, all of which engage the hormones that prompt the responses throughout our bodies that give rise to the mind/body symptoms we've discussed.

The logic behind cognitive restructuring is that, by learning to recognize, challenge, and change our negative thinking at the very beginning of this destructive cycle, we short-circuit that feedback loop, keep ourselves open and responsive to a more realistic, positive view of the world, and halt the development of stress-based illnesses.

■ I Think; Therefore, I Am Stressed

Like Kurt in the scenario above, we've all experienced situations that appeared overwhelming at first but later turned out to be much less serious than we thought. Sometimes, they prove to be the best thing that ever happened to us. In looking back objectively at our reactions, we realize that our initial thoughts were overly negative and out of proportion to the actual facts. That happened because our internal monologue had conjured up many more threats than actually existed.

The TM's mental chatter shifts continually from the past to the future, from hopes to fears, running through endless arguments and schemes. Through the TM's internal monologue, we mull over our needs and desires; create mental lists of what we have to do, or what we regret that we haven't accomplished. We endlessly review our worries and concerns, compounding and intensifying them while creating an exaggerated sense of time pressure to complete the tasks that face us.

This internal monologue only adds to our problems because it:

• Creates mental noise that distracts us and prevents us from "being" in the moment. Instead of experiencing the world directly through the Ancestral Mind, we live in a state of detachment and abstraction. We become alienated from our own immediate experience, losing the sense of vibrancy and authenticity in everyday things.

• Focuses on differences between ourselves and others, and between ourselves and the environment, and makes unbalanced evaluations of these differences. In other words, we might be quite comfortable living on $60,000 a year, until we discover that our old roommate is making $160,000 a year. The TM's monologue then intensifies all our unfulfilled expectations, our experience of loneliness, unfairness, and injustice, and our general frustration with life, all of which lead to the even more destructive feelings of guilt and anxiety. The negative emotions can overwhelm the mind and impair our ability to concentrate, perform daily activities, or motivate ourselves.

• Triggers stress responses that can brew below the surface of awareness. This unconscious mental and physical arousal creates a predisposition toward stress, which makes us even more vulnerable to other negative thoughts and emotions. We get trapped in a cycle of negativity and stress responses that becomes habitual. This cycle impairs our mental functioning and adversely shapes our view of the world.

• Prevents us from having a more grounded sense of self.

Our internal monologue is automatic and occurs largely outside of awareness, which is why we don't realize the effect it has on us. When we are faced with a stressful situation, the internal monologue assumes "tunnel vision," focusing on the problem and becoming preoccupied with it. All too easily, we get into the habit of falling back on what cognitive therapists call *negative automatic thoughts* (NATs).

In someone about to take a major exam, for example, negative automatic thoughts might be:

- "I hate this stuff."
- "I'm dreading this."
- "What if I fail?"
- "I know I'm not prepared."

If you get a message that your boss wants to see you, you might react with negative automatic thoughts like:

- "Oh, no, not again!"
- "What'd I do wrong now?"
- "Why me?"
- "Am I going to lose my job?"

Negative automatic thoughts typically involve worst-case scenarios, jumping to conclusions, "awfulizing," and "catastrophizing."[1] These NATs seem so real that we don't stop to question or examine them and eventually come to believe them as if they were fact-based, absolute truths. By accepting them as such, we lose our perspective and flexibility. We get locked into one-track thinking. We accentuate the negative, and view stressful situations in an inaccurate, skewed fashion.

Dr. David Burns of the Presbyterian Medical Center of Philadelphia, a pioneer in cognitive restructuring techniques, likens this filtering effect of NATs to a strong pair of eyeglasses that distort our view of the world and allow only the negative to pass through.[2] What gets filtered out, of course, are the positives.

From an evolutionary perspective, this type of filtering may be rooted in the same survival instinct we spoke of earlier. The trouble is, tunnel vision is no longer helpful for the kinds of threats we face today. In prehistoric times, when going up against some fierce predator, our ancestors were more likely to survive if they focused solely on the threat at hand. They also needed to respond quickly and automatically without conscious deliberation. In a world where saber-toothed tigers ran wild, it paid to assume the worst and respond accordingly. Those who looked on the bright side when they heard a rustle in the bushes often didn't make it back for dinner. They *became* dinner.

As we explained in chapter two, the control center for this kind of reaction is the amygdala, which sacrifices accuracy for speed. In sending along messages about what it perceives as immediate threats, it inevitably lets through a high number of false alarms. As the negative monologue responds not only to real stresses but also to imagined ones, it, too, triggers the amygdala to unleash alarm emotions such as anxiety and anger. While these responses might have been useful in defending yourself against a wild dog, they undermine health and well-being in a suburban home or a corporate office, when all they do is continue the feedback loop for stronger and stronger stress responses, particularly when they occur in response to stresses that don't really exist in the first place.

Obviously, not all negative thoughts are harmful in this respect. Grieving, for example, is part of the human condition, and sadness after the death of a loved one is entirely appropriate. Sometimes pessimistic thoughts help us to be careful and minimize risks, and angry thoughts can mobilize us to confront an injustice. The question we must ask ourselves is, are we responding honestly to an event, in the moment, or is our "alarm" response stuck in an endless loop of negativity in reaction to the TM's imagined scenarios?

There is also a significant degree of variability in different people's responses to stress—what we might call "emotional style." Not everyone would react to losing a job the way Kurt did. An ambitious, highly optimistic person might face such news with "You know . . . this is just the nudge I needed to find something better."

We have seen that genetics play a role in our emotional styles. We know that individuals differ in the relative activity of the left and right frontal lobes of the brain, areas that correlate with qualities such as optimism and pessimism. However, environment is also significant, for individuals raised in more nurturing, less stressful settings are less likely to exhibit negative emotional styles. Some of us are alarmists who see catastrophes in everything; others who are more optimistic can ride out stressful events without much fuss, or can at least turn off the negative monologue more quickly.

■ The Scientific Evidence for Cognitive Restructuring

Regardless of how strongly genetic and environmental influences affect our emotional style, we can learn to use cognitive restructuring (CR) to exert greater control over our negative mental monologue.

CR techniques were first introduced over thirty years ago by Dr. Aaron Beck, a psychiatrist, as a means of treating depression.[3] Since that time, numerous studies have shown CR to be more effective in helping severely depressed patients than either antidepressant medication or psychotherapy.

Quite notably, the improvements following CR hold up well at long-term follow-ups ranging up to several years, whereas pharmacological treatment does not. (Only 30 percent of depressed patients suffer a relapse of depression after CR, compared to 60 percent after pharmacological treatment.)[4] These findings indicate that the reason patients don't relapse after CR is that they've acquired a skill—changing negative self-talk—that they can use without relying on the intercession of a doctor or a medication. Drugs may relieve depression, but they do not produce enduring changes in the underlying cognitive process—NATs—that contributes to it.

CR also has been used successfully in the treatment of anxiety, panic disorder, phobias, obsessive-compulsive disorder, and post-traumatic stress disorder. As with its success in depression, CR has been shown to be more effective than psychotherapeutic and pharmacological treatment of many of these disorders. Several studies have demonstrated that learning to use CR allows patients who take medication for these problems to reduce their dosages.[5]

CR has likewise shown positive results with eating disorders and drug abuse and in reducing stress in patients with cancer, chronic pain, headaches, irritable bowel syndrome, infertility, cardiovascular disease, insomnia, and even HIV. Other studies have revealed that CR is effective in the treatment of chronic pain, arthritis, and Type A behavior—all conditions in which negative self-talk makes psychological or physical problems worse.[6]

But you don't have to have a specific medical diagnosis to benefit from cognitive restructuring. In everyday life, CR can enhance our emotional health by helping us to quiet the TM and reconnect to the Ancestral Mind.

Although no controlled studies have measured the effects of CR using brain imaging techniques, we suspect that it acts on the prefrontal cortex and working memory. As we discussed in chapter two, the PFC is the part of the Thinking Mind most responsible for self-consciousness, the internal monologue, and self-control. Given that cognitive restructuring reformulates the internal monologue, we can speculate that it exerts its therapeutic effects by altering neural activity and the contents of working memory in the PFC. We do know that excessive activation of the right side of this region has been linked to negative emotions, so it may be that, more specifically, CR minimizes negative emotions by dampening excessive activity in the right prefrontal cortex and allowing us to exert greater control over the thoughts we hold in working memory.

■ Minding the Mind

As we've said, CR can't change the situations that cause stress, but it can affect our emotional response to stress. The process begins with developing the ability to *recognize* our own cognitive responses. By identifying and stepping back from negative feelings, rather than getting caught up in them, we are better able to take a further step back to discern the internal monologue that is the source of those feelings. Once we *recognize* our own internal monologue, we can then *challenge* and *change* it. This means broadening the mental filter we bring into play when we experience stress and trying to retain our view of the bigger picture.

This process is not the same thing as simply thinking positively, and it is certainly not a matter of denying that a negative situation exists. Rather, it's a question of turning off the distraction pouring out of our Thinking Minds and allowing ourselves to perceive a situation more accurately. This requires a degree of reliance on emotional wisdom gleaned from past experience, and it's here that the Ancestral Mind can be especially helpful. By freeing ourselves from the tyranny of the Thinking Mind, we can ground ourselves in the more realistic Ancestral Mind, which deals only with what *is*, not with what the TM conjures up. Through millions of years of evolution it learned to filter out extraneous noise and focus on only those signals from the environment that were truly critical.

Once CR enables us to reduce the frequency and intensity of negative emotional reactions, and then to turn off those reactions more rapidly, it can eventually help us prevent them from occurring in the first place.

■ Eavesdropping on the Internal Monologue

The first step in cognitive restructuring is to become more conscious of your reactions when you confront stressful situations. This is not an easy task, however, for negative automatic thoughts are hard to recognize precisely because they are so "automatic."

At the Mind/Body Clinic at Boston's Beth Israel Deaconess Medical Center, we ask patients to track, in writing, the precise feelings and negative au-

tomatic thoughts that they experience when facing stress. Anyone can benefit from a daily Cognitive Restructuring Diary, which is nothing more than a
simple log with spaces for recording the most basic details:

COGNITIVE RESTRUCTURING DIARY

Situation   Emotion   NAT   Reframed Thought

By writing down, side by side, the details of a stressful situation, the emotion
that accompanied the situation, and the negative automatic thoughts that filtered through your mind and aroused the emotion, you will bring front and
center the fleeting but persistent NATs that usually exist only in the periph-
ery of your consciousness. You'll also become aware of just how often you engage in NATs. With this knowledge, you can begin to objectively examine
and weigh these thoughts, observing the distortions and inaccuracies they
give rise to and the negative emotions they perpetuate.

Here's an example of how the diary works. One patient who came to the
clinic was Andrew, a television news producer who suffered from anxiety,
anger, headaches, and insomnia. He routinely took sleeping pills to get him
through his sixty-hour workweeks. Part of our job in treating Andrew was to
help him learn about his internal monologue and the role it played in generating negative emotions and stress-related symptoms.

He began to track his thoughts with the Cognitive Restructuring Diary,
and after only a few days was able to identify the NATs that triggered his
anxiety and anger. He noted, for instance, that when he encountered a delay
while commuting to work, his thinking typically followed set patterns:

> "This road construction's been going on for two weeks—it'll never
>   end!"
> "I know I'm going to be late for my meeting!"
> "That driver cut me off. She's not going to get away with it!"

Before a meeting with his supervisor, he noted himself worrying:

> "This meeting's not going to go well."
> "She thinks I'm doing a lousy job."
> "She's not going to give me a performance increase."

Andrew found that charting his NATs on the Cognitive Restructuring Diary was an invaluable exercise, for he had never truly distanced himself from his habitual thinking to examine how frequently he engaged in NATs and how distorted they truly were. He also recognized how his perception of, and cognitive reaction to, stress was as critical as the stressing event itself.

■ Three Methods for Reframing the Negative Monologue

After identifying NATs and becoming aware of the feelings they arouse, the next step in cognitive restructuring is to reframe those NATs by formulating more accurate, adaptive thoughts about stressful situations and recording them in the diary, under the heading "Reframed Thought."

Reframing thoughts that have become habits takes some diligent work; however, there are a number of proven techniques that will make this easier for you. One technique is to ask yourself the following ten questions:

1. Is this thought really, literally true?
2. Am I overemphasizing a negative aspect of this situation?
3. What is the worst thing that could happen under the circumstances?
4. Is there anything that might be positive about this situation?
5. Am I "catastrophizing," "awfulizing," jumping to conclusions, and assuming a negative outcome?
6. How do I know this situation will turn out as I fear?
7. Is there another way to look at this situation?
8. What difference will this problem make next week, month, or year?
9. If I had one month to live, how important would it be?
10. Am I prone to using words like "never," "always," "worst," "terrible," or "horrible" to describe the situation?

These ten reframing questions are essential in guiding you to an understanding of the inaccuracies and distortions that color your negative automatic thoughts. They can help you realize that your NATs are not objective, fact-based assessments but only one of many ways of interpreting a stressful situation. Ultimately, they will lead you toward substituting a more accurate analysis of what you face in everyday life—one grounded in objective reality, without the filter of the TM's pessimistic, unbalanced babble.

The second reframing technique is to apply the "double standard" technique. Developed by Dr. David Burns, this method is based on the observation that, when it comes to explaining adverse events, we are often much harder on ourselves than we are on our friends. While we tend to have realistic and fair standards that we apply to others whom we care about, encouraging them to reframe their reactions to negative events more positively, by contrast we set unrealistic standards when trying to make sense of stressful events in our own lives.

To use the double standard technique, examine your NATs and then ask yourself: "Would I say this to a close friend with a similar problem? If not, what *would* I say to him or her?"

The underlying idea here is to reframe NATs by allowing yourself the same encouraging, empathic messages of support that you would give a friend. After all, don't you deserve at least the same degree of warmth and consideration you'd show to someone else?

The third technique is to weigh the current situation against past experience and ask yourself: "Has anything like this happened to me in the past, and, if so, how did it turn out?"

By asking this question, we can often prove to ourselves that many of the potential outcomes we worry about never do in fact happen. Or, if they do, they really don't turn out as badly as we fear. Once again, it's the AM that offers the necessary perspective: No matter what you're going through, the Ancestral Mind, grounded in the experiences of millions of years of evolution, can provide the emotional wisdom you need to endure and prevail.

Lauren, an actress, was also a patient at the clinic. Her life was much more under control than Andrew's—she was happily married, ate right, got plenty of exercise. The problem that brought her to us was the inability to sleep just before or just after performing.

One night, after appearing in a play for which she'd just received an enthusiastic round of applause, she met with her family and friends, who likewise offered their congratulations. One family member, however, couldn't restrain himself and suggested that Lauren "could have been better" in one scene. Instantly, Lauren's perception of her entire performance changed.

That night, Lauren noted her NATs in her Cognitive Restructuring Diary that made her feel anxious:

"The play didn't go as well as I thought."
"I'm just not making progress as an actress."
"What else did I mess up?"

By seeing her negative thoughts in writing and stepping back from them, she was able to begin to see how unrealistic they were. Then she used the ten reframing questions and wrote down this revised account in her diary:

"I overemphasized one comment from one person."
"I jumped to conclusions; even if one scene wasn't great, that doesn't mean the rest of my performance wasn't good."
"No performance is ever perfect."
"I couldn't have performed that poorly: I received long applause, congratulations, and praise from everyone else."

After reframing her NATs, Lauren could sense a change in her mood. She felt less anxious and more optimistic about her work and anticipated the next night's show. Happily, she knew she also would be able to look forward to a good night's sleep.

■ The "Stop-Breathe-Reframe" Technique

Once you gain experience recognizing and reframing NATs using the Cognitive Restructuring Diary, you are ready for the final step in cognitive restructuring: recognizing and reframing NATs at the very point that they filter through your mind using the Stop-Breathe-Reframe technique.

Stop

When you encounter a stressful situation, pay immediate attention to the NATs that you are experiencing, and then say "Stop" to yourself before the NATs escalate. A decisive act as simple as saying "Stop" can help break the cycle of NATs and the consequent undesirable emotional responses.

Because we're often more aware of the feelings that result from NATs than the NATs themselves, pay particular attention to your feelings and physical reactions when you encounter stress. These can be powerful cues that can help you recognize the cognitive miscues that distort your perceptions.

Breathe

After saying "Stop," take a deep breath. This will aid relaxation, divert your attention away from the Thinking Mind's cascade of thoughts, and, ultimately, help break the cycle of NATs and negative emotions.

Reframe

Reframe your negative thoughts using one of the three techniques described above:

- the ten key reframing questions
- the double-standard technique
- reflecting on past experience.

I vividly remember my first professional presentation to a large group of people. I had just accepted a position in a stress management clinic shortly after college graduation and had been asked to give a talk to a business luncheon. My hands were clammy and my stomach was in a knot. On that occasion, these were the NATs I recognized:

"I wish I didn't have to do this."
"I've never spoken in front of this many people before."
"What if I talk too fast and look rushed?"
"What if I forget parts of my talk?"

Fortunately, I had just learned the Stop-Breathe-Reframe technique as part of my training. I told myself to stop and take a few deep breaths. Next, I reframed my NATs by using the ten reframing questions, with which I was able to gain a broader perspective on the situation:

"I know this material and I'm well prepared."
"Actually, a little stress is good because it will energize my performance."
"I know I'll relax as soon as I begin my talk."

After using the Stop-Breathe-Reframe technique I felt much less anxious, and when it was time for me to make my presentation, I walked up to the podium, took a few deep breaths, and started in. I spoke slowly and confidently and, after a moment or two, found myself enjoying the experience.

Because NATs are so automatic and habitual, the Stop-Breathe-Reframe technique takes some time to master. But, with practice, you can learn to use Stop-Breathe-Reframe anytime and anywhere to turn off the negative stress filter, to catch and reframe NATs, and to develop a greater sense of control over your mental responses to stress.

Lauren, the actress, reported back to us that, while cognitive restructuring had not eliminated all her stress around performance time, it left her with an amount that was actually helpful. Like an athlete, she found that a manageable amount of anxiety enabled her to make the most of her talents. A little stress helped her to focus and be at the top of her game. She had returned to a more balanced response, reacting to stress only in proportion to the reality of the situation—the reality that, in its wisdom, the Ancestral Mind perceives. That's the reality that we, too, can focus on, once we learn to tame the Thinking Mind's distortions and distractions.

Chapter Seven

───────────────────────────────

The Power of Stress-Reducing Attitudes and Beliefs

Despite the anxiety-provoking intrusions of the modern world, and despite the distractions created by the Thinking Mind, some people cope far better with stress than others, and as a result, stay far healthier. How are these individuals able to maintain composure and well-being when so many others find themselves overwhelmed by their lives? They do so by maintaining an instinctive connection to their Ancestral Mind, at least in the form of the adaptive attitudes and beliefs that guided human evolution.

Qualities such as optimism and altruism aren't abstract principles concocted by the Thinking Mind, but are products of the Ancestral Mind and are grounded in the emotions. They persisted in the face of the pressures of natural selection because they helped early humans survive. Even in modern life, these beliefs remain closer to feelings, rather than abstractions, and when we let them, they motivate us toward thoughts and behaviors that improve our well-being.[1]

■ Optimism

Take a moment to consider the following questions:

> 1. When you think about the past, do you recall successes more easily than failures?

2. Do you usually attribute positive outcomes to yourself instead of luck?
3. When you think about the future, do you expect the best?
4. Do you believe that something positive can be found in most negative situations?
5. Do you feel that things usually don't work out for the best for you?
6. Do you find it hard to count on good things happening to you?

If you answered yes to the first four questions and no to the last two, you are likely an optimist; if you said no to the first four questions and yes to the last two, you are probably a pessimist. The point of this exercise isn't to assign a label to your character, however, but to get you to begin to consider how a tendency toward optimism or pessimism can have a significant effect on your mental and physical health.

Optimism is the feeling that, despite frustrations and setbacks, things will turn out for the best. Optimists, by nature, feel secure about themselves, the world, and the future. They focus on and expect positive experiences, are more likely to attribute positive outcomes to themselves, and believe they can influence events through their actions. Optimists also expect the best when faced with uncertainty, view setbacks as temporary, and remember successes better than failures.

While optimists certainly experience the full range of thoughts and emotions, their attitudes and beliefs create a mental filter that allows in primarily positive thoughts while blocking negative ones. Optimists are not bothered by as many negative thoughts, and they can change from negative to positive thinking more easily. They also enjoy more positive moods and greater feelings of self-esteem and well-being than pessimists.

In our discussion of the physiology of the AM, we saw that greater activity in the left prefrontal cortex is associated with a positive emotional style, including optimism, while greater right prefrontal activation heightens negative thoughts, beliefs, and emotions. Optimists may be able to dampen negative thoughts and emotions due to their ability to "turn on" the left prefrontal cortex under stress; pessimists, in contrast, have a harder time turning off the right prefrontal cortex. Because optimists can regulate the kind of information that the thalamus and reticular formation send to the neocortex, they allow mainly positive thoughts to enter conscious awareness and working memory and can prevent negative thoughts from doing so.

Psychologist Richard Davidson, who has conducted the most extensive research on the relationship between frontal brain asymmetry and emotional style, has found that there is a consistent tendency for greater left prefrontal activation in the population as a whole.[2] This apparent bias toward optimism in the neural circuitry of the brain supports the argument for its adaptive benefits.

Anthropologist Lionel Tiger believes that optimism is not only inherent in human nature but is one of our most defining and adaptive characteristics.[3] He speculates, however, that optimism developed in humans as a counterbalance to one particular aspect of the Thinking Mind—namely, our ability to think ahead about the future. As we know all too well, when we contemplate the future, we are prone to begin to imagine dire consequences, including our own mortality.

I believe that optimism provided a more fundamental evolutionary advantage: It played a crucial role in the pursuit of goals, motivation, and persistence in the face of challenges and adversity. Confidence that you can succeed at difficult tasks increases the likelihood that you will attend to, rather than avoid, such tasks, and that you will have the perseverance required in order to succeed. If you have faith in a positive future, then current tribulations and sacrifices may appear bearable, even essential. Because many of our technological achievements required tremendous motivation, steadfastness, and trust in progress, optimism likely continued to have significant adaptive value even as the Thinking Mind was increasing its influence over our behavior. Because optimism improves mood and motivation, it also adaptively buffered us against apathy, hopelessness, and chronic depression (chronic depression would have been maladaptive, as it impairs impulses crucial to survival, such as appetite and sexual desire). Through its mood-enhancing effect, optimism also may have facilitated social interaction and prosocial behaviors such as cooperation and sharing.

On the other hand, unwavering optimism is not always a desirable trait. A reasonable amount of pessimism forces us to be prudent and minimize risks, and it heightens realistic, accurate perceptions of danger. Emboldened by unbounded optimism, an early human who challenged a raging lion single-handed would soon be culled from the gene pool. Assuming that you can make it home safely while speeding on icy roads, or that you can continue to smoke indefinitely without developing lung cancer, are modern equivalents in which a dose of pessimism could enhance survival.

94 The Ancestral Mind

True pessimists, in contrast, hold persistent negative beliefs about themselves, circumstances that affect them, and the future, and are more likely to view negative events in the worst possible light. They interpret negative incidents

- personally (they blame themselves for the events)
- permanently (they see bad events as never-ending and think in terms of "always" and "never")
- pervasively (they believe negative events will undermine everything).

Thus, a pessimist who fails to get hired for a job may conclude:

"It's all my fault."
"I'll never get the job I'm looking for."
"I'll never be successful in life."

It is easy to see how this kind of thinking can become a self-fulfilling prophecy. It is a maladaptive intrusion of the Thinking Mind, which impairs the confidence, enthusiasm, and self-esteem that are necessary to succeed at almost any task.

Pessimists also tend to encounter more bad events in life, in part because they are less proactive in avoiding them and more passive in reacting to them when they do occur.[4] Pessimists tend to be struck by more illness, because negative beliefs induce stress, which impairs health.[5] Pessimists also have weaker social support systems, which is a further risk factor for increased morbidity and mortality. Pessimists may encounter more slips, fender benders, and household mishaps because their characteristic bad moods may distract them or lead to more risky behaviors.[6] This is especially problematic in that such negative events can then trigger even more unhealthy stress reactions.

Optimism is linked to overall happiness and achievement, and actually may increase life expectancy.[7] It improves health, increases the likelihood of healthy-behavior practices, reduces susceptibility to disease, and enhances recovery from disease. Optimism also has been linked to good morale; perseverance and effective problem-solving; academic, athletic, military, occupational, and political success; popularity; and freedom from trauma.[8] One study found that a pessimistic personality style predicted which college students would report a large number of sick days and visits to a physician.[9] Another study found that, of men who suffered their first heart attack, optimists were significantly more likely than pessimists to be alive almost ten years after the heart attack. In this study, optimism was a stronger predictor of survival than the actual amount of heart damage or coronary blockage, or factors such as blood pressure or cholesterol level.[10]

In a major study involving members of the Harvard University classes of 1939 through 1944, men were rated on levels of optimism and pessimism at age twenty-five, then tracked for thirty years. Overall, the men who were pessimistic at age twenty-five were less healthy and had more chronic illness later in life than the men who were optimistic.[11]

Optimism has also been associated with fewer physical symptoms such as headaches; it also predicts success in alcohol treatment programs, college students' performance (a study involving five hundred freshmen at the University of Pennsylvania showed that optimism was a better predictor of grades than SAT scores or high school grades, probably because it is a good indicator of motivation), insurance agents' sales performance, and presidential election victories.[12]

Other studies show that optimistic people recover faster from surgery, have fewer medical complications during and after surgery, and exhibit stronger measures of immune functioning than pessimists.[13] In a study conducted at UCLA on immune parameters in first-semester law students, researchers found that optimism was associated with higher numbers of helper and natural killer cells.[14]

While optimism may be due largely to inborn temperament, nurturance also plays a role. Children who experience stressful life events such as divorce or death are more prone to pessimism. Similarly, children may learn early in life to mimic their parents' emotional styles. Recent studies indicate that children of parents who are negative, critical, and blaming are more likely to exhibit pessimism, as are children whose parents discipline them by inducing guilt or temporarily withdrawing affection.[15]

The good news, though, is that no matter what the source of their emotional predilection, people can learn to become more optimistic. We have already discussed one prime technique for promoting optimism: cognitive restructuring. This method can enable you to recognize, challenge, and change pessimistic feelings, quiet the negative self-monologue, and think with greater cognitive flexibility. By learning to broaden your perspectives and interpretations of situations, cognitive restructuring can help you to reframe pessimistic thoughts into more optimistic ones, to see the glass as half-full instead of half-empty. The Cognitive Restructuring Diary is a valuable tool for identifying situations that trigger pessimistic thoughts, which can then be challenged and reframed by using the various exercises we explored earlier.

A number of the other mind/body techniques that we will investigate likewise promote optimistic thinking by dampening negative emotions and activating positive thoughts, feelings, moods, and behaviors. You will learn that one technique in particular, visual imagery, can be very effective in helping you to think optimistically by imagining and mentally rehearsing positive outcomes.

You can also promote optimistic thinking and beliefs by using a mental checklist involving the "Three Ps" to reorient yourself when you encounter adversity or negative situations:

Personal

Permanent

Pervasive

Personal means to avoid blaming yourself for negative events, particularly those that are beyond your control. This idea was stated beautifully by the late American theologian Reinhold Niebuhr, who wrote, "God grant me the serenity to accept the things I cannot change, the courage to change the things I can, and the wisdom to know the difference." Accepting personal responsibility is logical only when you're dealing with what you can truly influence; things like the actions of others are not realistically under our control.

Permanent means to avoid interpreting a single negative event as part of a never-ending downward spiral. If you receive a poor grade on an exam, don't automatically assume that you will perform poorly on subsequent exams. Similarly, a job interview that doesn't lead to an offer should not cause you to conclude that you will never land a job. See the future in terms of what you *can* do instead of what you *can't* do.

Pervasive means to avoid generalizing a problem so that it applies to your entire life. If a relationship doesn't work out, don't tell yourself that your relationships always turn sour. If you throw a party that flops, don't tell yourself, "I can't do anything right."

You can also turn the principles of the Three Ps around and use them to view positive events in optimistic ways. For example, when you have a good experience on the job, *do* view it as permanent ("Things tend to go well for me professionally most of the time"), pervasive ("Lots of things in my work life are going well"), and personal ("This success isn't just temporary or due to luck or external causes, but happened because of my abilities"). Attributing positive outcomes to your own abilities will strengthen your self-confidence, which in turn will promote greater self-esteem and more optimistic moods.

Another technique for strengthening optimistic thinking is optimistic affirmation—repeating positive beliefs to yourself regularly:

> "I expect the best when faced with uncertainty."
> "For every obstacle, there is a solution."
> "I can handle this."
> "I'm doing the best I can."

Call upon optimistic affirmations when encountering adversity, thinking about the future, or before going to sleep. Over time, these positive statements will become more automatic and unconscious than the pessimistic beliefs to which you've become accustomed.

Here are three other ideas for strengthening optimistic thinking:

• *Practice an "attitude of gratitude."* Optimists amplify positive thinking through appreciation. If you focus on what you have and on positive events in your daily life, the negatives you must face don't loom so large. Instead of fretting about what you lack and your shortcomings and difficulties, consider those things that you appreciate: good health, family, friends, work, an event you look forward to, or something as simple as what someone did for you today. We can all find many positives in the course of the day if we look for them.

• *Limit your exposure to the media.* Radio, TV, and newspapers saturate our minds with pessimistic stories about the dangers that lurk around every corner. While being a good citizen may demand that we keep up with certain issues in our community and around the world, we really have no need to absorb every detail of every outrage, calamity, or celebrity peccadillo.

• *Avoid pessimists and seek out optimists.* Optimism and pessimism are both contagious. The next time you're thinking negatively, spend time with a young child. Children's boundless optimism is inspiring and infectious; their excitement is fuel for positive thinking. Even the mere laughter of a child can make the most pessimistic person smile.

Optimism is not just a matter of healthier thinking; it concerns the power to choose one's attitudes. Things turn out the best for people who make the best of the way things turn out. You didn't have a choice about being an optimist or a pessimist when you were a child, so don't blame yourself for what you have inherited by way of genetics or upbringing. But now you do have a choice. You can choose to be an optimist, and with enough determination, you can make that choice a reality.

■ Anger: The Toxic Emotion

In our survey of ancestral life, we saw that hostility and aggression were far less important to human development than altruism and cooperation. However, anger is an entirely appropriate and adaptive response when one is being hurt, threatened, treated unjustly, demeaned, or blocked from realizing an important goal. Throughout human history, anger has inspired humans to right injustices, to mobilize others to change inappropriate behavior, and to keep motivated to stick with difficult tasks until they were accomplished.

The trouble is that, for some people, anger occurs too frequently and persists too long. Excessive anger is not adaptive. It does not lead to positive behavioral change but becomes counterproductive. It disrupts relationships, and only serves to heighten stress. Anger also intrudes on the Thinking Mind, impairing our ability to think clearly, to concentrate, and to perform effectively.

When anger becomes chronic it can also affect health, particularly when it takes the form of hostility, a more intense, pervasive type of anger that involves cynicism, animosity, and aggression. Because hostile people come to expect that others will mistreat them, they mistrust others in advance, seeing everyone as the enemy. They live with a permanently short "fuse" and an overreactive amygdala, which causes blood pressure and heart rate to rise, blood fat and cholesterol to increase, blood platelets to become "stickier" so that they block artery walls, blood vessels to constrict, and oxygen flow to the heart to decrease.[16] When these conditions persist, they can lead to serious illness, even death.

Scores of studies document that people with high levels of anger and hostility are at greater risk for heart attacks and heart disease. Research also shows that reducing anger and hostility diminishes the risk of recurrent heart attack and may in fact prevent heart trouble.[17]

Chronic anger and hostility also weaken the immune system, with adverse affects for our overall health. Dr. Redford Williams of Duke University followed a group of hostile men from about age twenty-five to age fifty and discovered that they were seven times more likely than less hostile men to die prematurely. Williams found that anger was a stronger predictor of early death than smoking, high blood pressure, and elevated cholesterol.[18]

Another way in which hostility adversely affects health is that angry people tend to drive friends and family away. Because they mistrust and show a lack of empathy for others, angry individuals are more likely to reject the help of others. As a result, they do not avail themselves of the beneficial effects of social support, and thus place their health in further jeopardy.

A series of studies has revealed that hostile people report fewer, less satisfactory social supports in their personal and professional lives, as well as greater conflict and less satisfaction with marriage.[19] Although it is possible that hostile people are angry because they have a poorer social network, it is more likely that the reverse is true. In a long-term study of 150 Swedish men, hostile men who were socially isolated had three times the death rate of those who were less isolated;[20] even if someone is hostile, therefore, an adequate social support can help.

One hallmark of the hostile individual is excessive self-involvement and selfishness. He is far more likely to use the words "I" and "me" rather than "we" and "us."[21] Whereas children and emotionally healthy adults who are in touch with their Ancestral Mind have an expanded sense of ego boundaries that allows them to feel a sense of closeness and trust with others, hostile people have constricted ego boundaries that cause them to emphasize their differences with other people.[22]

Hostile individuals view others with suspicion and regard their behaviors as irritating or as failing to live up to expectations; they have a hard time congratulating others, are preoccupied with the errors and mistakes of others, and are quick to argue.[23] They exhibit a pervasive need to defend themselves from "the enemy"—other people—who pose a constant threat. As a result, hostile people feel isolated and are never at peace with themselves.

We are more prone to anger when we are stressed, irritated, or frustrated by something else; this "shortened fuse" can be the result of unconscious physiological arousal that accumulates below the level of awareness. In other words, if someone is upset because the alarm didn't go off and is running late for an early meeting, he is more susceptible to anger if the children are difficult during breakfast or a traffic jam occurs on the way to work. The pervasive stress of modern life can create unconscious arousal, which is more likely to make people feel hostile or angry all the time. While men and women get angry equally often and just as intensely, men are more likely to externalize their anger by yelling, throwing things, or slamming doors.

Fortunately chronic anger, like relentless pessimism, can be changed. The process takes considerable practice, but we can learn to lower the threshold for anger and turn it off more easily. At least eighteen studies have demonstrated that hostility can be reduced.[24] Furthermore, eight controlled behavior-modification programs designed to reduce hostility in heart attack victims have shown an average reduction in recurrent heart attacks of 39 percent and a 33 percent reduction in deaths when compared with standard cardiac rehabilitation programs that don't focus on reducing hostility.[25]

Eleven Exercises for Reducing Anger

Exercise #1

Acknowledge that your anger is often problematic. You can't come to terms with anger until you admit that it's a behavior that you want to change. Tell someone that you realize you have difficulty with anger. Your open acknowledgment, along with the support of others, will maximize your success. And ask yourself whether you really enjoy anger and the loss of control that it often involves. By recognizing that many anger-related behaviors—exploding, sulking, using hurtful words or acting foolishly—are unpleasant, you will be more motivated to live a different way.

Exercise #2

Cognitive restructuring techniques can help reduce anger by reframing angry thoughts. Pay particular attention to the Cognitive Restructuring Diary described in the previous chapter; it can be used to identify patterns that cause anger, whether times and places, particular circumstances or people, and so on. By becoming more aware of the frequency and kinds of situations that trigger anger, you will be better able to recognize, anticipate, and control it. You may also see how trivial some of the events that provoke your anger really are.

Exercise #3

Practice the Stop-Breathe-Reframe technique described in the previous chapter. It will help you develop a greater ability to control and turn off anger as it happens.

Exercise #4

Practice the relaxation response and visual imagery techniques that we will explore in subsequent chapters.They have been proven not only to reduce anger by quieting angry thoughts, but also to produce carryover effects that minimize stress and unconscious physiological arousal. Consequently, our fuse isn't as short and our threshold for anger is higher. Practice all of the other mind/body methods that we will discuss. Since anger and other negative emotions go hand in hand, these techniques will decrease anger by improving mood, heightening positive emotions and dampening negative emotions, and dissipating unconscious physiological arousal. Many of the skills diminish an unhealthy, excessive focus on the self, which is the root cause of cynicism and hostility. When you are less caught up in yourself, you will be better able to engage in prosocial behaviors, and it will be more pleasant for others to be around you.

Exercise #5

Don't expect perfection. Be realistic and modify your expectations for the behavior of those around you. Given that a hallmark of the hostile individual is excessive self-involvement, become more aware of how frequently you use the terms "I" and "me."

Exercise #6

In anger-producing situations, cool off by taking a time-out, using distraction, taking a walk or drive, or engaging in a pleasurable pastime. Such activities quiet anger by preventing us from focusing on negative thoughts.

Exercise #7

Exercise regularly and avoid excessive caffeine or alcohol. As we will learn, exercise is an outlet for tension and produces a tranquilizing effect that reduces anger. Caffeine is a stimulant that makes us more susceptible to anger by heightening tension and physiological arousal. Because alcohol is a depressant that reduces self-control, it, too, can lower the threshold for anger.

Exercise #8

Try to develop a keener sense of empathy. Reframe anger-producing interactions from the other person's perspective by imagining his or her feelings. Being a good listener and developing an attitude that promises, "I want to understand what you are feeling," are cornerstones of empathy.

Exercise #9

Pay more attention to your social support network. The comfort and caring of others can help you reframe and short-circuit anger.

Exercise #10

Follow the common spiritual admonition to forgive when you feel you have been wronged, and treat others as you would have them treat you. Blame, resentment, and revenge feed the angry heart. One reason religions stress these principles may be that they have known for centuries what science has only recently discovered—anger undermines health and may even have deadly effects.

Exercise #11

Put anger in perspective by asking yourself how important an anger-producing situation would be if you had only one week to live. Many people who have had someone close to them die unexpectedly can place anger in its proper context more easily. They step back, take a broader look at their life, and realize that many of the things that make them angry are, in reality, insignificant. They realize that love and support are all that really matter.

■ Laugh Your Stress Away

How often do you laugh?

Common sense tells us that laughter reduces stress, anxiety, anger, and depression by allowing us to shift to a less intense and rigid perspective on ourselves and our lives. Learning to see the humor in a stressful situation not only provides a time-out from the pressure but facilitates cognitive restructuring, both by reframing irrational thoughts and by enabling a more positive view. A humorous outlook doesn't make stress go away, but it does allow us to disengage from it, maintain a healthy distance, and see the bigger picture. With a new viewpoint, we sometimes find fresh insights and original solutions to our problems.

Science has in fact documented that strong laughter numbs pain, produces a mild state of euphoria, and is thought to release endorphins, the brain's opiate-like chemicals that generate a natural "high."[26] In one study, subjects were measured on their sense of humor, then exposed to a variety of stressors. People with a low sense of humor showed a greater suppression of immune functioning in response to stress, suggesting that a better sense of humor provides an effective buffer.[27] Humor also has been shown to moderate the effects of stressful life events on mood disorders such as depression. It also increases people's tolerance for higher levels of physical discomfort.[28] In *Anatomy of an Illness*, the late Norman Cousins wrote an account of how he cured himself of a severe arthritic condition by using Marx Brothers films and *Candid Camera* episodes to stimulate daily laughter. He found that ten minutes of hard laughter had an anesthetic effect that reduced his pain and improved his sleep.

Being able to laugh at ourselves is particularly stress-reducing, because it helps us to put our own shortcomings in context. Whatever dumb thing we may think we have done, we can see that it probably was no dumber than the errors made by a thousand people before us.

Aside from being a great tension release in itself, laughter reduces barriers between people. By increasing a sense of connectedness, it likewise increases positive emotions.

We tend to associate easy laughter—most often in the form of uncontrollable giggles—with the freedom of childhood. Unfortunately, many of us lose touch with our sense of humor as we age. Some adults have been so stressed and serious for so long that they've almost forgotten how to laugh.

Fortunately, reestablishing a connection to the humor around us is a stress-reduction technique that can be learned. Some of the simple prescriptions that follow may seem painfully obvious, but with so many grim and troubled faces out there, maybe even these simple and obvious pointers aren't being "taken seriously."

Rx #1: Actively seek out humor as if it were a prescribed daily medicine.

Take time to read your favorite comic strip in the newspapers. Rent comedies at the video store. If you're not a fan of the type of humor that Hollywood has to offer these days, you can still find the Marx Brothers, or classics like *Some Like It Hot* or *Mr. Roberts*, which might just take your mind off your problems. When you come across a funny article in a magazine or newspaper, save it and find someone to share the laugh with.

Rx #2: Make it a point to avoid other people who are too serious.

Actively cultivate people who might make you laugh, even if you find you may not have anything else in common. Funny people make everyone around them feel better about themselves. When you hear a good joke, write it down. Maybe you'll find the wherewithal to tell it to others.

Rx #3: Pay a visit to the bookstore.

There are hundreds of desk calendars available—*The Far Side* and *Garfield* come to mind—that allow you to begin every day with a joke. Search out the humor section, or ask a salesperson for suggestions for novelty books and lighthearted novels.

Rx #4: Remember that even adults are allowed to be playful.

When was the last time you jumped on a trampoline or built a snowman? Activities like these can make us laugh and feel young again. Make a point of watching how children enjoy themselves. A happy child, living in the moment, is a great teacher for getting in touch with the Ancestral Mind.

Rx #5: If you have to at first, fake it.

Research shows that merely by "putting on a smiling face" we can actually change our thoughts and make ourselves feel better. (By the same token, a frown can reinforce sadness.) So pay attention to your expressions. Every time you stand in front of a mirror, practice your smile until it becomes a natural response.

Rx #6: Laugh at yourself.

People who take themselves less seriously, with a self-deprecating sense of humor, have greater self-esteem. Exaggerate if you have to in order to put your stressors in perspective. David Sobel and Robert Ornstein, authors of the *Healthy Mind, Healthy Body Handbook*, use the following example of a daughter at college writing this letter home to her parents:

> Dear Mom and Dad:
>
> I am sorry that I have not written, but all my stationery was destroyed when the dorm burnt down. The car crash that followed when we drove away wasn't as bad as it seemed at the time, for we were all alive. I am now out of the hospital and the doctor said that I will be fully recovered within a few years, and I may well be able to walk one day. I have also moved in with the boy who rescued me, since most of my things were destroyed in the fire.
>
> Love, Hilary
>
> P.S. There was no fire, no accident and my health is perfectly fine. In fact, I do not even have a boyfriend. However, I did get a D in French and a C in math and chemistry, and I just wanted to make sure that you keep it all in perspective.

■ Using Positive Illusions and Denial to Manage Stress

Psychologists have found that stress-resistant people are more than just optimistic: They actually distort reality in order to view it in the best possible light.

Psychologist Shelley Taylor has conducted extensive research at UCLA on stress-resistant individuals[29] and identified three factors she calls "positive illusions": mildly distorted positive beliefs about themselves, an exaggerated confidence in their ability to control what goes on around them, and unrealistically optimistic notions about the future. To describe this set of beliefs, which appears to be especially important when people are faced with threatening information or stressful events, Taylor coined the term "positive illusions."

In one study, she found that many breast cancer patients with the highest degree of adjustment to their condition believed they had a great deal of personal control over the cancer and were overly optimistic in their assessments of their odds of survival, despite evidence to the contrary. In another study on men with AIDS, those who indicated that they had a realistic view of their disease's course died an average of nine months sooner than those who were more unrealistically optimistic about their disease.[30]

Ironically, the traditional definition of mental health is characterized by a worldview that is in close contact with reality, and such "reality testing" became the hallmark of sanity. However, in the 1970s evidence began to show that most people, in fact, are neither truly realistic nor accurate in how they think. Memory is an area in which people are particularly inclined to refashion the truth to suit their own purposes.

Taylor's research took these findings one step further: she suggested that distorting reality in a mild way is actually *healthful* and that a lack of positive illusions is correlated with depressed or highly anxious individuals. Those who are mildly depressed, in fact, see themselves, the world, and their future far more *realistically*. Severely depressed individuals, however, clearly suffer from negative, distorted views of themselves, the world, and the future that are directly proportional to the severity of their depression. Perhaps, as Taylor points out, there is a lesson to be learned from the depressed patient who asked his psychiatrist, "What's so great about reality, anyway?"

The beneficial aspects of powerful illusions extend directly to our attempts to heal our physical ailments. Ancestral man recognized that the brain has its own internal pharmacy that can be activated through belief. For millennia, sacred and mystical healers—whether medicine men, shamans, or witch doctors—have relied upon the power of belief to induce therapeutic mind/body interactions to aid healing. Using special clothes, rituals, and incantations, they create an atmosphere of mystery and authority to heighten their healing abilities. Witch doctors also use the paralyzing force of negative beliefs in the practice of voodoo, casting spells on an intended victim. Many instances of voodoo death have been reported in the medical literature, concerning people who, as the result of a major negative emotional experience such as fright, suddenly die. In the 1940s the renowned Harvard physiologist Walter Cannon studied voodoo death in the Maori aborigines of New Zealand through their use of *tapu*, or taboo. Cannon described *tapu* when used by the tribal chiefs against others as "a fatal power of the imagination working through unmitigated terror."[31]

In Western medicine the power of belief has been appreciated since the time of Hippocrates. "Some patients," he said, "though conscious that their condition is perilous, recover their health simply through their contentment with the goodness of the physician." One hundred fifty years into the modern era, the Greek physician Galen observed, "He cures most in whom most are confident."

Prior to the 1900s, most medicines given by physicians had little or no curative power, yet patients still improved. The explanation for this phenomenon is what is called the placebo effect: the patient's and physician's belief in the efficacy of a medical treatment, which mobilizes strong self-healing mechanisms. Today, when newly developed medications are scientifically evaluated, they are compared to an inactive placebo pill in a "double-blind" study in which subjects and researchers don't know whether a particular individual has received the new medication or the placebo. (Eventually, the researchers unseal a code that informs them which pill each subject received.) Numerous studies have shown that, for virtually all health problems, including anxiety, depression, pain, fever, headaches, high blood pressure, angina, acne, asthma, insomnia, ulcers, and arthritis, approximately one-third of patients improve when given a placebo.[32] Likewise, about one-third of patients in pain will respond as well to a placebo as they will to morphine, which is the most powerful narcotic ever developed.[33]

The placebo effect can even reverse the action of potent drugs. In a widely cited case study conducted by Dr. Stewart Wolf in the 1950s, a woman who suffered nausea and vomiting during pregnancy was given a medication which she was told was "new" and "effective" and would quickly relieve her symptoms. Within twenty minutes, the woman's nausea and vomiting stopped even though she had, in fact, taken ipecac, a medication that induces vomiting.[34]

While the mind (and body) can benefit by occasionally suspending a rigid adherence to hard-nosed reality, a problematic variation on "selective" attachments to reality occurs when people negate stressful events by resorting to denial. It is essential to realistically confront major traumas, health problems, and other significant stressors. Research on stress-resistant individuals indicates, however, that denial of minor, everyday problems actually can be healthy. Minimizing or ignoring minor problems—the small insults in life that have no real significance or enduring implications—certainly makes life more manageable. As we discussed in chapter two, the ability to prevent information from reaching consciousness involves the reticular formation and thalamus, the two principal brain mechanisms in the AM that regulate attention and awareness. Stress-resistant individuals have, in some way, learned to exert greater control over the neural activity of these brain structures. Specifically, they are able to shut down the flow of information from these areas to the cortex so that stressful information never reaches conscious awareness.

Renowned stress theorist Richard Lazarus has termed this kind of denial "positive denial." He has shown that a refusal to believe in something negative, as long as it is not carried to an extreme, is associated with well-being in the face of adversity. Positive denial helps keep hope and morale up and reduces anxiety.[35] In contrast to positive denial, being preoccupied with minor problems only serves to increase stress and its deleterious effects on mood and well-being.

Even though positive illusions can cause people to ignore or deny important information, make bad decisions, or develop a potentially harmful false sense of their abilities or accomplishments, Shelley Taylor maintains that positive illusions are generally subject to feedback from the world, in the form of both failed personal actions and the reactions of other people. These reality tests usually keep illusions from becoming too extreme or maladaptive. Furthermore, individuals typically experience "time-outs" from their positive illusions, particularly when setting goals and making decisions. Once those decisions and goals are established, the illusions are reactivated to help them carry through their plans. Occasional periods of mild depression or melancholy may also allow time-outs from positive illusions to keep them within healthy limits. During these periods, people slow down, readjust their appraisals, and tone down their behavior and decisions.

As Taylor is careful to note, however, many stressors in life are not amenable to change through positive illusions. And even when stressors are amenable to such change, positive illusions can only help us to cope, without removing the stressors altogether.

Positive illusions also contribute to positive mood and prosocial behaviors, such as the ability to care for others, both of which are health-enhancing. It makes sense that individuals who handle stress more effectively, experience more positive moods, and are more optimistic will have a more favorable view of their fellow man, react to others more enthusiastically, and have a stronger sense of empathy. In part this is because stress-resistant people are less distressed about themselves, and in part because happy people tend to attract better social support systems, simply because they're more pleasant to be around.

For the healthy, stress-resistant person with the Thinking Mind and the Ancestral Mind in good balance, beliefs are characterized by an adaptively positive bias toward reality. Given the tendency toward greater left frontal brain activation and positive affective style in the population as a whole, it may be that, just as the neural circuitry of the brain is inclined toward optimism, it is also inclined toward positive illusions.

Positive illusions are present to an even stronger degree in children, where they mirror the animism of ancestral man: the belief that inanimate objects have feelings and can interact with us. Young children think more positively of themselves; believe they are capable of magically modifying reality (such as being able to control the movements of the sun and the clouds);[36] see themselves as popular; and have very positive beliefs about the future. These illusions contribute to an expansion of the boundaries of the self and a sense of communion with others and the world. In short, they comprise the same sense of wonder and mystery that, in part, defines the ancestral state of mind.

Taylor believes that positive illusions may be especially effective during childhood in facilitating the acquisition of language, problem-solving abilities, and motor skills. The unrealistic and optimistic beliefs of children may ensure that youngsters persist at learning during the first few years of life, minimize responsiveness to criticism, and enhance self-esteem and motivation. So once again, positive illusions appear as an evolutionary accommodation wired into the brain—in this case, essential for both cognitive development and emotional health.

As we "grow up" and away from the AM, entering TM-based adulthood and its greater absorption with rationalism, we become more responsive to realistic limits placed on our abilities, and we "unlearn" the illusions of childhood. But as positive illusions diminish in strength, so, too, do a sense of mystery and a more unified sense of self. One of the unfortunate consequences of the passage to adulthood is the diminution of positive illusions and the loss of the magical cosmos of childhood. The most stress-resistant individuals may be those who, by keeping the Thinking Mind's rationalism in a healthy balance, and by retaining a greater degree of conviction in their positive illusions, also retain a stronger, more integrated sense of self.

■ The Power and Biology of Faith

As we've seen in the example of the shaman, medicine has been intertwined with sacred and spiritual beliefs throughout history. In ancient times, doctors were priestlike figures who were considered to have power over the spirits. Although religion and science have diverged in modern times, and often seem at odds, several dozen major studies have begun to document that religious faith and spirituality may contribute to the reduction of stress and the improvement of health. Consider the following:[37]

• Those with religious beliefs have lower rates of suicide, drug use, juvenile delinquency, divorce, and morbidity and mortality. Overall, at least six studies in the past several years have consistently found that religious involvement is associated with a 25 percent reduction in mortality, which is equivalent to a seven-year difference in survival.

• Patients with religious beliefs are more likely to survive longer after cardiac surgery than those without religious beliefs.

• Religious and spiritual beliefs offer some protection from stress-related illnesses, including hypertension, heart disease, stroke, and cancer.

• Religious and spiritual beliefs are associated with reduced anxiety, anger, and depression, and with greater life and marital satisfaction, intact and stable families, well-being, and self-esteem. In fact, in a review of over one hundred studies that measured religiosity and well-being, seventy-nine of the studies reported a relationship between religious involvement and greater happiness, life satisfaction, and morale or positive mood.

• The greater the religious involvement of those of advanced age, the less their physical disability.

• People who attend religious services have lower blood pressure than, and half the risk of heart attack, of nonattendees. Religious beliefs may be as important a factor in heart disease as traditional risk factors such as cholesterol and smoking.

• Several studies have discovered that individuals with greater religious involvement exhibit stronger immune systems.

• Religious beliefs are associated with fewer unhealthy behaviors, such as smoking and drinking.

• Religiously active people cope better with bereavement, divorce, unemployment, and serious illness.[38] In one study, for example, patients with advanced cancer who found comfort from their spiritual beliefs were more satisfied with their lives, happier, and had diminished pain.[39] Religiously active people are happier, and for the elderly, one of the best predictors of life satisfaction is religiousness.[40]

When spiritual beliefs instill fear and are repressive and controlling, they can, of course, become a negative influence, and some religious groups also avoid contact with health care professionals, reject life-saving medications or medical procedures, and fail to vaccinate their children. However, the incidence of such tenets is small, and the scientific evidence for religion's positive effects on health is far more compelling.

It may be that some of the positive health effects of religious beliefs can be attributed to what used to be called "clean living." Religion also usually provides a rich social support network, promoting volunteer activities and community life, and reducing loneliness and isolation. Dr. Herbert Benson of Harvard Medical School maintains that religious beliefs and prayer may be health-enhancing because they elicit the relaxation response, an inborn quieting mechanism that we will explore further in chapter nine. He suggests that when we pray, we enter a peaceful state that enhances feelings of joy and contentment. Because the relaxation response improves health, he believes the tendency for humans to engage in religious beliefs and prayer may be encoded in the physiology of the Ancestral Mind. Benson contends that we are "wired for God"; that is, we have always known that believing in a higher power was good for us. On a more fundamental level, this belief also counteracts our fears about our own mortality.[41]

Nonetheless, religious affiliation and spiritual beliefs also foster many of the other adaptive and stress-reducing attitudes and beliefs we associate with the Ancestral Mind: optimism; a sense of control, commitment to something outside of oneself, and challenge; forgiveness and tolerance of the imperfections in oneself and others; empathy and altruism.

By giving us a broader perspective on our lives, spirituality prevents our being caught up in the materialism, individualism, and smaller daily frustrations that surround us. When crises occur, they can be interpreted through a spiritual lens as necessary steps to growth. Religious beliefs help us make sense of tragedy and suffering. They reduce stress by providing a sense of meaning, hope, and purpose, as well as a philosophical system that allows us to organize our experience in a way that is more intelligible and coherent. Through spirituality, we gain perspective on our connection to others and the world. Above all, religious belief satisfies the need to know that we matter.

Ultimately, religious beliefs are one of the most deeply personal aspects of our existence, and what works for one person won't necessarily work for another. But if only for the sake of your emotional and physical health, try opening up your receptivity to the spiritual dimension of life, then exploring a path to faith with which you feel comfortable.

Chapter Eight

Social Support and Stress Hardiness

During the time that AT&T underwent the largest corporate reorganiza-
tion in history, Dr. Suzanne Kobasa studied the company's business executives. In this period of high stress and uncertainty, she observed that some
executives stayed healthy while others did not.

Kobasa discovered that the stress-resistant executives possessed a uni-
form feeling about themselves and their lives, which she termed "stress hardiness." She defined the stress-hardy personality as being characterized by
three beliefs:

- Control
- Commitment
- Challenge[1]

Stress-hardy individuals have a sense of control over events in their lives, a
strong commitment to something outside of themselves, and an ability to
view stress and change as challenges and opportunities instead of threats.

By contrast, Kobasa found that the AT&T executives who became ill felt
powerless, threatened, and debilitated by change and uncertainty. They retreated from stress and exhibited a sense of alienation from others. They preferred stability, and, as a result, were more likely to find life boring and
meaningless.

Because stress-hardy people are committed to something larger than themselves, they take better advantage of social support, and are deeply involved in both their work and their families. Just as it did in ancestral times, this connectedness provides a sense of meaning and direction in their lives. Stress-hardy people view change as normal and challenging. They are curious about their environment and interested in new experiences, even seemingly stressful ones, which they view as integral to growth and personal transformation. They also are optimistic by nature.

Of the three components of stress hardiness, it may be that the most important is control. Studies have demonstrated that, with a sense of control—even the illusion of control—people can tolerate extreme stress. Those who feel a high sense of control rather than a belief in fate or luck, or a conviction of their own helplessness, not only cope better with stress, but do better in life and live more happily, by a number of important measures.

In one study, a group of students was given a series of math problems to complete, during which they were subjected to random bursts of loud noise. The researchers told half the students that they could stop the noise by pressing a button, even though, in reality, the button would do nothing. This segment exhibited fewer physiological signs of stress, even though none of them ever pushed the button. The mere suggestion that they could control the stress kept them calmer.[2] In a similar study in which some students could control noise during math problems while others could not, the students who had no control were shown to have higher levels of stress hormones in their blood.[3]

Individuals exposed to stress that they view as uncontrollable, such as the care of a relative with a debilitating illness like Alzheimer's disease, have been shown to exhibit poorer immunological functioning; they are also at greater risk for developing major depression.[4]

Research by psychologist Martin Seligman and his colleagues suggests that when animals and people experience uncontrollable events, they internalize the message that their actions have no effect on outcomes, and they become passive and unresponsive. This "learned helplessness" is a major risk factor for a number of illnesses.[5] In one study, pairs of monkeys were confined to a chair for eight hours a day. One of the monkeys in each pair had to turn a light off once a minute to prevent an electric shock from being delivered to the tails of both monkeys. The monkeys who had to turn off the light developed high blood pressure, but two-thirds of the monkeys with no control over the light died from cardiac arrhythmias.[6]

Perhaps the most striking study of the effects of sense of control on health was conducted by psychologists Ellen Langer and Judith Rodin at Yale University. Their research involved changing the amount of control that elderly nursing home residents had over daily events in their lives.[7] On one floor of the home, residents were given choices over variables such as what movies to watch, which plants to take care of, the types of eggs they could eat for breakfast, and so on. Another floor was given no choice in any aspect of their daily routine. Eighteen months later Langer and Rodin conducted a follow-up assessment of both floors and found that the residents on the floor given more control were more active, happier, and, most impressive, fewer had died. The sense of control not only improved well-being, but actually saved lives.

Belief in control over events in the environment is also associated with better health habits. If you trust in the value of exercise, for instance, you are more likely to get to the gym or take off on that brisk walk, which will indeed improve your health. This credence in control, called self-efficacy, is central to efforts to modify many behaviors such as diet and nutrition, and smoking.

All of the techniques in this book, but particularly cognitive restructuring and the relaxation response, maximize a sense of control over your mind, body, and stress. Over time, this sense of control will evolve into an almost unconscious belief that you have greater control over stress than it has over you.

Just as you can learn to become more optimistic, you can develop a sense of stress hardiness by:

• Telling yourself optimistic stories about events in your life and practicing the techniques we explored in the previous chapter for developing optimism.

• Viewing your life as meaningful, problems as opportunities, and the future as a challenge.

• Believing that you have some control over your environment and your responses to it.

• Developing a deep source of commitment to something other than yourself.

• Viewing change as normal, inevitable, and as a stimulating, healthy challenge, not a threat.

■ **Help Yourself by Helping Others**

Altruism is the feeling, developed through natural selection, that most directly connects us to a cause beyond ourselves.

Too much preoccupation with ourselves can lead to anxiety and depression by increasing concentration on problems; altruism reduces the focus on ourselves and serves as a distraction from worries.

Itself a product of the ancestral way of life, altruism nourishes our connection to the Ancestral Mind by providing these other benefits:

- Improved positive emotions such as caring
- Improved attitudes and feelings of greater contentment with what we have
- Increased self-esteem and sense of well-being by strengthening belief in our own skills and strengths
- Reduction in anger and social isolation and bolstering of social support.

Allan Luks, author of *The Healing Power of Doing Good*, surveyed thousands of volunteers to document a phenomenon called "helper's high," which results from practicing altruism on a regular basis.[8] "Helper's high" consists of sensations of warmth, increased energy, and euphoria, and can lead to long-term relaxation and calm. Harvard Medical School psychiatrist George Vaillant, who followed Harvard graduates for forty years, identified altruism as one of the major qualities that helped graduates cope with the stress of life.[9]

Helping others doesn't just feel good, and it doesn't only give life greater meaning. Research shows that altruism can reduce the effects of stress and dramatically improve the health of the helper. In one major study of 2,700 residents in Tecumseh, Michigan, men who volunteered for community organizations were two and a half times less likely to die (from any cause) than men who did not volunteer.[10] According to Drs. David Sobel and Robert Ornstein, authors of *The Healthy Mind, Healthy Body Handbook*, studies also indicate that helping is associated with boosted immune functioning, fewer colds and headaches, and relief from pain and insomnia.[11]

An altruistic attitude in daily life can be fostered by participating in many activities: tutoring, telephone work, visiting nursing homes, helping at hospitals or homeless shelters, cooking or delivering meals, or donating money.

Some of these activities can be done alone; others may involve working with a community organization. The important thing is to choose an activity that you enjoy doing and with which you are comfortable.

A regular donation to your favorite charity is a great idea, but keep in mind that there may be some benefit to choosing activities that require personal contact and that allow you to share directly in the responses of the person you are helping. Also, try to choose something that you are good at, since, obviously, these are the activities you'll find more personally rewarding.

Even simple, spontaneous acts in daily life can be altruistic: holding the door open for someone, assisting an elderly person, or helping someone whose arms are full; letting another proceed ahead of you in line or in traffic; shoveling a neighbor's driveway after a snowstorm; or sending a note or making a quick phone call to let someone know he or she did a good job.

And one more strategy for enhancing altruism in your life: Practice the other mind/body techniques in this book, which, by amplifying positive emotions, will make it easier to engage in altruistic behaviors.

■ No Man Is an Island

Do you have people you can talk to when you're troubled? Do you feel supported and cared for by other people, or do you feel isolated and lonely?

A predisposition to depend on social support is rooted in the Ancestral Mind as a genetic memory, and as a basic emotional need. Our ancestors were neither fleet-footed nor equipped with natural weapons such as claws, so in order to protect themselves from more physically powerful species and to hunt and gather food, they relied on a network of close social relationships. Adults who formed such attachments were more likely to come together as a group to nurture their offspring, and social bonds kept children close to their caregivers so that they were protected from harm.

Social support begins with the attachment of child to parent. Humans have the longest infancy in the animal kingdom, which makes social connections to caregivers a matter of life and death. Studies also consistently demonstrate that infants' mental and physical development are directly linked to adequate attachment and nurturing.

Unfortunately the conditions of TM-based modern life disrupt the AM's biologically based need for social support by causing an increase in social isolation and lack of connectedness. Throughout evolution man lived in extended kin networks, surrounded by genetic relatives such as aunts and uncles, nephews and nieces. In the past several decades, however, social ties have been weakened by mobility, fragmentation of the nuclear and extended family, single-parent families, and rising rates of separation and divorce.[12] Few of us have parents who live nearby, and few enjoy the same close-knit communities that existed in previous generations. Lifelong friendships are not as common. Most of us don't just drop over to a friend's house unannounced, as was the case in years past. As a population, we are more likely to live alone and remain unmarried, and far less likely to belong to a social organization.[13]

Substantial evidence indicates that people with adequate social support—defined as family, friends, community contacts, social or religious organizations, work relationships, or even a pet—are healthier, less likely to develop major and minor illnesses and mental health problems, are better able to resist communicable diseases, have a lower death rate, and cope better with various stressors.[14] Specifically, studies show that social support provides this wide array of benefits:

- Reduced susceptibility to diseases of all types, from cancer to arthritis to heart disease
- Diminished risk of depression, stress-related illnesses, and premature death
- Lowered cholesterol and incidence of coronary heart disease
- Better coping with bereavement, job loss, and illness.

When serious illness does strike, individuals with social support do better medically, recover more quickly, and have lower death rates.[15] In a review of the literature spanning several decades and almost forty thousand people, researchers concluded that lack of social support doubles the risk of morbidity and mortality.[16]

Another major study found that mortality was inversely related to each of these measures of social involvement: marriage, close friends and relatives, church membership, and informal group associations.[17] And, in a landmark study on women with breast cancer conducted at Stanford University, researchers found that those who participated in social support groups lived twice as long as those who did not.[18] Consider these other findings on social support:

• In a 1997 study conducted by researchers at Carnegie-Mellon University, volunteers were intentionally dosed with a virus. The researchers found that the more diverse the subject's social relationships, the less likely he or she was to catch a cold.[19]

• In a ten-year study, single men in the age group forty-five to fifty-four die at twice the rate of married men.[20]

• In a three-year study of over two thousand men, socially isolated survivors of heart attacks were more than twice as likely to die earlier than those who were less isolated.[21]

• Medical students under academic stress who are lonely show greater suppression of immune functioning than students with adequate support.[22]

The negative effects of social isolation hold true regardless of race, ethnic background, sex, age, or socioeconomic status.[23] In fact, inadequate social support is as detrimental to health as lack of exercise or high cholesterol; it is as great or greater a risk factor for death than smoking![24]

Of course, isolation is not the same as solitude, for many people who live on their own are happy and healthy. It is, rather, the subjective sense of being isolated and alone, cut off from people and having no one to confide in, that is problematic. By the same token, not all social ties are health-enhancing. Some negative relationships, most especially abusive ones, can be stress-inducing and have been associated with increased illness and suppression of immune functioning. In one study, marital conflict was found to weaken the immune response of both newlyweds and couples who had been happily married for decades.[25]

Nonetheless, there remains a significant advantage for those who have strong, stable connections to others. Some studies show that merely being married, whether satisfactorily or not, offers more protection from illness than being single. In fact, a man who marries can expect to automatically add about nine years to his life.[26] Widowers have been found to have death rates from three to five times higher than married men of the same age for every cause of death, and the recent death of a spouse is associated with higher death rates during the first six to twelve months of bereavement.[27]

Married people not only have a regular source of social support in their spouse; they also are better integrated into the community as a result of being married.

Some of the most surprising findings on social support come from studies on the Japanese. Even though they exhibit a high incidence of nicotine use and hypertension, consume a high-fat diet, endure elevated levels of stress that include six-day workweeks, and live in polluted and crowded cities, they have one of the lowest rates of heart disease and enjoy among the longest life expectancies in the world.

Researchers believe a primary reason that the Japanese are not subject to the increased morbidity and mortality that would normally accompany their lifestyle is their significant level of social support. Japanese culture, in contrast to most Westernized societies, is group oriented, with a high commitment to family and community. The Japanese emphasize social stability and stronger social ties than Americans do; they also honor and respect the elderly and place a high priority on lifelong friendships.[28]

Social support provides many important health-enhancing benefits: love and affection, empathy, social activity, camaraderie, a sense of purpose and belonging. As a kind of safety net that we can fall back on during times of stress, it serves to help suppress negative thoughts and emotions when people's view of themselves, the world, and the future are threatened. Social support can also enable us to feel good about ourselves and enhance our sense of control and optimism when we're under stress—when self-esteem, optimism, and belief in control are likely to suffer.

The positive emotions resulting from social support, including love, contentment, and warmth, engage health-enhancing circuits in the AM; the stress-reducing effects minimize excessive activation of its circuitry of negative emotions. Social support also exerts positive effects on health because it allows us to:

- Share feelings
- Seek solutions to problems with the help of others
- Reframe negative thoughts and change behavior
- Develop a sense of reliability and stability in times of transition
- Shift attention away from ourselves to something larger
- Access physical or material assistance.

The act of confiding in others may be one of the most important benefits of social support. In over two decades of research on the advantages of revealing personal pent-up thoughts and feelings, Dr. James Pennebaker of Southern Methodist University has demonstrated that confiding can be liberating, health-enhancing, and produce particular physiological effects. Pennebaker has shown that merely writing about emotional experiences can boost immune functioning, resulting in fewer physical symptoms and doctor visits.[29]

Even one of the basic foundations of family life that has traditionally served as a regular and dependable time of intimacy, the nightly family meal, is fast disappearing in our overscheduled, hectic lives. The family meal has always functioned as a time to talk about the day, share feelings, and listen to one another's successes and problems. But with both parents working and the incredible number of activities in which our children participate, the family has a harder time getting together regularly for dinner. Eating together has always been a kind of home base for families, one that strengthens a sense of security and belonging; when that base is eroded, so, too, is family cohesion.

In the end, one of the great ironies of modern life may be that, despite our being surrounded by many more people than our distant ancestors were, we have fewer intimate relationships and more loneliness and isolation than at any other time period in human history. Here are some steps you can take to avoid social isolation:

• Commit to the belief that "people need people." You will reduce stress and its deleterious effects on health, boost the immune system, and probably live longer.

• Use cognitive restructuring and the other mind/body techniques mentioned in this book, which will help improve your mood. You will find it easier to develop a social support network when you are in a better frame of mind.

• Strengthen your social support network by joining groups, clubs, or community organizations that interest you. Check the local newspaper, community events calendar, colleges, or religious organizations for relevant information. Sign up for a class on cooking, interior design, or photography.

• Instead of sitting home alone in your spare time watching television, call someone on the telephone, visit a neighbor, meet someone for a cup of coffee, shopping, dinner, or a movie. Take the initiative. Make the effort to nurture your friendships. Remember that it is the quality, not just the quantity, of your relationships that is important. The best way to attract friends is to be a friend.

• Avoid the tendency to withdraw under stress. Instead, seek out family and friends and make an effort to socialize. Be sure that you have at least one confidant, and preferably more, on whom you can depend in life.

• If you're not allergic, get a pet. Animals can be a source of social support, affection, and good company. Studies show that pets lower the blood pressure of their owners and that pet owners have fewer doctor visits, lowered illness rates, higher survival rates for heart disease, and longer lives.[30] Pets provide a chance for interaction with a living thing, fulfill our innate desire to care for, and be cared for, and alleviate loneliness.

In the next chapter, we're going to return briefly to the world of our ancient ancestors to learn more about our natural endowment for stress reduction, and what we can do to develop a most important stress-reduction technique: eliciting the relaxation response at will.

Chapter Nine

——■————————————————————————■——

Opening the Door to the AM:
The Relaxation Response

To explore the natural rhythms of sleep and wakefulness, Dr. Thomas Wehr of the Psychobiology Branch at the National Institutes of Health conducted a series of experiments attempting to replicate the experience of sleep as ancestral man might have known it.[1]

In setting up his study, Wehr tried to re-create conditions in the latitudes inhabited by early man, latitudes in which a midwinter night lasted for about fourteen hours. He surmised, based on anthropological evidence, that our ancestors would have remained in the safety of their caves, campsites, or sleeping bowers during that entire fourteen-hour period. Wehr therefore had his experimental subjects stay in darkness from 6:00 P.M. to 8:00 A.M. every night for an entire month. With no artificial light, schedules, or alarms, these subjects were able to enter a cycle of sleeping and waking based on nothing other than their own internal, physiological cues.

What Wehr found was that, during the fourteen hours of darkness, his subjects lay in a state of quiet rest for about two hours before falling asleep. Then they slept for about four hours, awakened from dream sleep into another two-hour period of quiet rest, and then fell asleep again for four hours more. Each morning, the subjects awoke around 6:00 A.M. from their dream sleep, then spent another two hours in quiet rest before arising at 8:00 A.M.

Wehr concluded that ancestral man, with no choice but to endure the darkness, and following no schedule but his own natural rhythms, slept for eight hours, as we do, but that his sleep, like that of the experimental subjects, was noncontinuous. It was interspersed with another six hours (in three two-hour blocks) spent in a state that was neither fully asleep nor fully alert and active, a state that Wehr described as "quiet rest with a physiology all its own."

This pattern is similar to the sleep routines of many mammals. Its technical name is *polyphasic sleep*, and it is the pattern we experience early and late in life. Babies sleep in multiple periods during the day, and older adults likewise tend to sleep noncontinuously at night and nap during the day. Evidently, this alternation met an evolutionary need.

Modern life, as dictated by the TM, has not acknowledged a role for this state of quiet wakefulness through much of the night, yet it lies dormant within our physiology, part of our genetic makeup. Instead, we use artificial light to stay active and remain awake well into the evening, then consolidate our sleep into one seven- to eight-hour block for most of our adult life.

What this sleep pattern achieves, however, is to cut us off from the significant periods of deep relaxation—up to six hours in the winter months—that our ancestors likely experienced throughout their evolution. Noting that these phases of quiet wakefulness usually occurred immediately after dream sleep, Wehr speculated:

> One wonders whether dream sleep might have had a greater impact on our ancestors. At some time each night they would have awakened from dream sleep and entered an extended period of quiet wakefulness in which the effects of dreams might reverberate in conscious awareness . . . It is tempting to speculate that in prehistoric times this arrangement provided a channel of communication between dreams and waking life that has gradually been closed off as humans compressed and consolidated their sleep. If so, then this alteration might provide a physiological explanation for the observation that modern humans seem to have lost touch with the wellspring of myths and fantasies.

Wehr describes "quiet wakefulness"—what we might call "daydreaming" or "reverie"—as a deeply relaxed, nonverbal, and more receptive mode of mental functioning, akin to what psychologists call "primary-process." This form of mental activity, at which the Ancestral Mind is so skillful, makes sense of information not by logical, formal rules but by greater conceptual flexibility and intuitive connections, which combine information into meaningful schemes via images and emotions.[2]

You've experienced reverie if you've ever lost yourself "drifting off" while lying on the couch, staring at breaking waves at the ocean, or gazing into the crackling logs of a fireplace. The consequent reduction in the normal barrage of sensory input into the AM relieves the "pressure" of sensory stimulation and makes unconscious emotional stimuli more accessible. Even if these stimuli are not experienced consciously (because our level of alertness is generally reduced), the reticular formation and thalamus are now less active and the release of emotional stimuli is cathartic. Deep relaxation dissipates the mental and physical tension caused by the unconscious emotional arousal that accumulates beneath the level of awareness and affects our behaviors, moods, and health. As a result, we restore energy and feel renewed and refreshed.[3]

More than a hundred years ago Freud devised his method of "free association of ideas" to enable the conscious mind to gain access to unconscious emotional information. He called it a "relaxation of the watch upon the gates of reason, the adoption of an attitude of uncritical observation." But free association as practiced in psychoanalysis is a process of the TM, a way of trying to understand and interpret the meaning of associative mental activity using the conscious mind.[4] It is not the truly relaxed, free-flowing mental state of reverie, which is a property of the Ancestral Mind alone.

When individuals enter the deep relaxation of reverie, they produce characteristic theta brain wave patterns (alpha EEG is a "relaxed wakefulness" brain wave; theta EEG is more common during states of deep relaxation, such as the transition from wakefulness to sleep). Because the TM "lets go" of its usual volitional control over conscious thought processes, individuals who are awakened from this state report an abundance of rich imagery, thought to represent the spontaneous emergence of unconscious feelings.[5] This is called *hypnagogic imagery*, and it is typically described as dreamy, drowsy, floating, wandering. Many have compared the experience to being a passive spectator at a movie or play. The imagery appears as disconnected "snapshots," as opposed to the usual form of dreams, which are characteristically longer and better organized.[6]

We experience hypnagogic imagery every night just before we fall asleep. When we close our eyes, we first spend a few minutes in a state of relaxed wakefulness that is characterized by drifting thoughts and alpha brain waves. Our thoughts gradually wander until the body begins to relax. We next pass through what's called *stage 1 sleep*, which is really the same thing as reverie—the drowsy, relaxed state between waking and sleeping.[7] Activity decreases in the reticular formation and thalamus, so our awareness dims. Our bodies become more relaxed: muscle tension lessens; respiration and heart rate decrease; slow, rolling eye movements occur; and theta brain waves are generated as alpha waves begin to disappear. Because we enter deeper stages of sleep after only a few minutes in stage 1, we typically don't remember the hypnagogic imagery that passes through our minds.

In our waking lives, deep relaxation, reverie, and hypnagogic imagery can become a source of creativity and an aid in problem-solving.[8] Many scientists, artists, and writers have reported the essential role of reverie in the initial stages of the creative process. Perhaps the most famous example is the chemist August Kekulé, who derived the idea of the benzene ring from a hypnagogic image of a snake biting its own tail that appeared to him in a dreamlike state. Mark Twain, Edgar Allan Poe, and Robert Louis Stevenson all incorporated into their work the results of spontaneous imagery summoned up during reverie.

Accessing the Ancestral Mind—which means suspending the operation of the TM—is valuable to problem-solving because in reverie, the intuitive mode of hypnagogic imagery yields combinations of ideas that precede the logical working-out of a proof. This is precisely the sort of mental playfulness we often call intuition or "a hunch," a playfulness that allows the organization of ideas into new structures.

■ The Relaxation Response

The benefits of deep relaxation are so profound that it has become a cornerstone of mind/body medicine as it has developed over the past thirty years. The physiology underlying every phenomenon we've just discussed—from our ancestors' hours of quiet rest, to creative reveries, to our stage 1 sleep each night—has been exhaustively studied under the rubric of "the relaxation response."[9] This research not only has revealed the specific details of how and why our bodies enter into this state, but also has enabled us to perfect the techniques that allow us to enter this state at will. The relaxation response is the AM's inborn calming mechanism, an adaptive response refined through millions of years of evolution. As a method of promoting health, it is safe, has been proven effective, and is absolutely free.

In earlier chapters we explored how stressful stimuli activate the circuitry of negative emotions in the amygdala, thalamus, and hypothalamus, triggering the "fight or flight" response, which includes a rise in levels of stress hormones, more rapid heart rate, greater muscle tension, and so on. We also discussed how all these stress reactions can occur completely outside the range of our awareness.

For most of the course of our evolution, the stress response was part of an adaptive balance of negative and positive emotions, designed to see us through life-threatening emergencies. In our natural habitat, the system included an automatic "shutoff": Once the stress response had elicited the burst of physical energy necessary to deal with a given threat, the Ancestral Mind would trigger the relaxation response to bring the body and mind back to a resting state.

In the world created by the Thinking Mind, the stressors we encounter are seldom acute, well-defined physical threats that require fight or flight. More often they are chronic and psychological, and the negative emotions are not subsequently balanced by positive ones. Because our stressors don't require a concentrated burst of energy, there is no natural dissipation of the physical arousal of the stress response. The stress response stays "switched on," which leaves us in a chronic state of mind/body arousal, with greater readiness to experience emotions like anxiety or hostility, and with myriad stress-related health problems.[10]

The key to confirming the existence of the relaxation response was biofeedback. Physiologists had always assumed that we had no voluntary control over the autonomic nervous system, the branch of the body that governs basic functions such as respiration and heart rate. But biofeedback experiments developed in the late 1960s changed those perceptions.[11] Researchers were now able to use electronic instrumentation to measure physiological data such as brain waves and heart rate, which were then fed back to a subject using auditory tones or visual signals. In this manner, an individual was given precise information about his or her own unconscious physiological processes.

With these methods scientists discovered that, by having subjects alter their mental activity (whether thoughts, images, concentration, or attention) and use the biofeedback signal as a "mirror," those subjects could change such "autonomic" functions as brain waves, heart rate, blood pressure, blood flow and skin temperature, and muscle tension. This discovery opened the door for exploration of techniques such as meditation, long claimed by adherents to provide precisely this kind of inner control.

It was Harvard Medical School cardiologist Herbert Benson who dubbed the underlying mechanism the relaxation response (RR). As Benson defined it, the components of the relaxation response include:

- slower brain wave patterns and mental quieting
- reduced secretion of stress hormones
- reductions in heart rate and breathing rate and, in some cases, blood pressure
- increased blood flow to the extremities
- relaxation of the muscles throughout the body.

Benson also pioneered the development of explicit techniques to make this form of deep relaxation readily achievable. He isolated four key elements to bring it about:

- a quiet place with eyes closed to minimize distractions
- a comfortable position and muscular relaxation
- a mental focusing device such as breathing, a word or phrase, or an image with which to shift the mind away from everyday thoughts
- passive disregard of everyday thoughts.

Research has proved that the RR is an effective treatment for a variety of health problems, including anxiety disorders and panic attacks; headaches, back pain, arthritis, cancer pain, and other chronic pain conditions; gastrointestinal problems, such as irritable bowel syndrome; hypertension, angina, and heart disease; menopausal hot flashes, premenstrual syndrome, and infertility; and nausea and vomiting from chemotherapy. The RR can also be used to stabilize blood sugar levels in diabetics and speed recovery from surgery. It is routinely employed to reduce the length of labor and discomfort of childbirth and has been shown to strengthen the immune system and increase defenses against upper respiratory infections.[12]

Although the relaxation response has become a standard treatment in mind/body medicine, until recently our understanding of its therapeutic effects has focused on its ability to counter the stress response in the body. But my own research on brain wave changes has demonstrated that the relaxation response also dampens neural activity in the frontal cortex, which is not only the site of the TM's self-monologue but also part of the AM's circuitry of emotions. In precise terms, it's the "beta EEG," an alertness brain wave, that decreases during the relaxation response. In other words, the RR may exert its therapeutic effects, in part, by "quieting" emotional activity and the internal monologue in the frontal lobes.[13]

When people practice the RR regularly, they can achieve deeper states of relaxation. Studies of RR techniques such as meditation have consistently found that, during the RR, brain waves slow as measured by heightened alpha and theta EEG activity. Alpha activity typically rises during the first few minutes of the RR, followed by increases in theta level. (Some studies have found that, if the individual exhibits high amounts of alpha activity prior to beginning the RR, then only theta activity increases during the RR.) Several researchers have established a direct relationship between the length of time subjects practiced the RR and the degree of EEG changes: Those with years of experience always exhibited increased theta activity during the RR.[14]

In one study, I used more precise, computerized analysis of EEG activity in the frontal and temporal (side) brain regions to assess the effects of the RR in subjects who had been practicing the RR for only eight weeks. I found that, compared to a control condition, theta activity increased during the RR while alpha activity decreased. This suggests that theta is, in reality, the key brain wave that changes during the RR. The ability to produce increased theta activity was a function of practice; the greater the number of weeks of practice the subjects practiced the RR, the greater their increase in theta activity during the RR.[15]

The appearance of slower brain wave patterns like theta waves indicates that the reticular formation and thalamus are reducing the arousal level of the cortex so that the cortex can relax, take a break from its normal processing of information, and conserve energy. The fact that regular practice of the RR resulted in increased theta waves suggests that the RR exerts its calming effects by quieting the brain.[16] (These findings are consistent with studies that have demonstrated that, because the resting state of the human brain is highly activating due to the energy required to inhibit the constant barrage of sensory input into the brain, focused attention deactivates brain regions such as the cortex.) These quieting effects are likely the basis for the positive feelings of well-being that are commonly reported for the RR and, ultimately, the reduction in stress responses that is associated with improvement in stress-related health problems.

In summary, then, the primary effect of the relaxation response is not just in quieting responses in the muscles or in the cardiovascular system, but in acting on some of the structures that make up the Ancestral Mind itself. Its effects include:

• Producing states of mental and physiological quiescence that diminish mental activity in the frontal cortex (the site of the TM's monologue). This effect is probably achieved by focusing attention on some repetitive stimulus that interrupts the TM's internal monologue.

• Lowering the AM's level of arousal by inhibiting the flow of sensory information from the reticular formation and thalamus, providing instead monotonous sensory stimulation. The constant stimulation of daily life bombards the AM endlessly. By directing attention inward, the RR helps us to close out the distractions of the outer world and achieve lower arousal states that quiet negative thoughts and emotions and heighten feelings of relaxation.

• Reducing physiological input from the body by relaxing the muscles. The AM receives ongoing feedback from the musculature about emotional processes, which can influence stress reactions in a cyclical fashion. By relaxing the muscles, the RR assists the AM to turn off the stress response.

• Diminishing sense of self and self-reflective awareness by reducing sensory input to the orientation association cortex and working memory.

Besides quieting the brain, the RR may have another fundamental therapeutic effect in opening the channels of communication with the AM by suppressing the TM's normal, rational thinking processes. This allows us to gain access to stressful emotional stimuli that have been registered unconsciously in the AM, as we saw was the case for reverie and primary-process states. The aim of many systems of meditation is, in fact, to permit the individual to gain awareness of feelings and emotional stimuli that have previously been inaccessible. Meditation frees our attention from distracting sensory stimuli in order to focus on the more subtle stimuli, particularly emotional ones, that unconsciously shape our perceptions and behavior.[17]

■ Siesta Time

The RR and our predilection for quiet rest may also be linked to our innate propensity to need an afternoon nap. Although many people believe the familiar slump in mood and performance in the early afternoon is the result of eating a heavy lunch, it actually occurs because the Ancestral Mind was programmed by evolution to perform better with mid-afternoon rest.

The napping of toddlers and the elderly, and the afternoon nap of siesta cultures, have led sleep researchers to the same conclusion: the brain intended that we take a nap in the middle of the day.[18] The tendency for grogginess in the afternoon is present even in good sleepers who are well rested. If we don't always notice this is true, it's probably because we're running around too fast or we've stoked ourselves up on caffeine.

Because the midday nap is an integral part of so many cultures, particularly those near the equator, the practice may be part of an evolutionary mechanism not just to reduce stress but to get us out of the hot midday sun. Nonetheless, at least one study has found that stress has the strongest negative effect on our mood and on our immune systems in the afternoon.[19]

The urge for a nap is not as strong as the need for nighttime sleep, and can therefore be suppressed. As naps conflict with work timetables, they are becoming less common. The irony here is that, because the rigid scheduling demands of the TM prohibit naps, they actually cause a decline not just in mood, but in energy, alertness, and performance.

Research suggests that an afternoon nap even as short as ten minutes can enhance all these factors, especially after a night of poor sleep. In one study, pilots on long-distance airplane flights who were allowed to take a break from the controls for a brief nap in the cockpit were less fatigued, more alert and vigilant, and performed far better. There is also some evidence that naps may have other health benefits. In a study conducted in Greece, daily naps were correlated with a 30 percent reduction in the incidence of coronary heart disease.

Other findings have shown that shorter naps of less than one hour comprise a significant amount of wakefulness and light sleep, including stage 1 sleep, so napping may be more similar to the RR than to the deep sleep of nighttime. But the benefits of a nap don't appear dependent upon reaching any specific stage of sleep. What is significant is attaining deep relaxation in the middle of the day to counter the typical afternoon surge in stress.

■ The Restorative Power of Sleep

While the previous sections have, I hope, demonstrated the value of naps and of learning the relaxation response, there's probably no substitute for a good night's sleep. After all, nature has seen to it that we spent at least one-third of our entire evolutionary history sound asleep.

Among the reasons that sleep has such an important impact on our health and well-being are:

• Deep sleep (which is characterized by delta brain waves, or what sleep researchers call slow-wave sleep) affords the Ancestral Mind a period of reduced blood flow and energy metabolism, which appears crucial for restoring energy and optimal neural functioning. Activity in the reticular formation, thalamus, limbic system, and prefrontal cortex is also significantly diminished.[20] Energy, joy, optimistic thinking, coping with negative emotions, and positive moods are all dependent on getting enough deep sleep.

• Some sleep researchers view dream sleep as the brain's system for maintaining emotional homeostasis by processing emotions, routinely accessing emotional memories, and regularly suppressing self-consciousness. The reticular formation, thalamus, and limbic system all become very active during dream sleep, with increased blood flow and energy metabolism, so dream sleep may be therapeutic to the AM in discharging emotional arousal.[21]

Nonetheless, and particularly in light of the stress of the TM's modern life, half of adults complain of disturbed sleep. Stress is the most common precipitator of insomnia, and many of us have a harder time sleeping on stressful days.[22] When you're stressed during the day, your stress hormones remain elevated even much later, when you're trying to sleep.

Dozens of scientific studies have shown that the RR is an effective treatment for insomnia. When practiced during the day, the RR counters stress responses, leading to lower levels of stress hormones at night. The RR also improves sleep itself because, when practiced at bedtime or after waking in the night, it can produce the brain wave pattern that is similar to stage 1 sleep. In other words, the relaxation response can take you to light sleep, which is your portal to deep sleep and dream sleep.

Throughout our evolution, the relaxation response has been our constant companion. It mitigates physical stress, and it calms us before and after sleep. It improves mood and energy and enables us to open the channels of communication with the AM. It is part of our genetic makeup and, as we'll explore later, is linked to our need for nature, solitude, and other calming stimuli. Reconnecting with the relaxation response is a readily available way to reestablish a link to with the Ancestral Mind. And it's not at all hard to do.

Learning to Elicit the RR

There are three simple steps to eliciting the RR:

Step 1.

Relax the muscles throughout the body. This is accomplished by allowing relaxation to spread gradually throughout the body. The feelings that result may vary from warmth, heaviness, or tingling to nothing specific.

Step 2.

Establish a relaxed breathing pattern. When we are in a state of calmness or sleeping, we breathe with the abdomen, which relaxes the body because carbon dioxide is expelled and oxygen inhaled efficiently. Under stress, we tend to breathe inefficiently, using short, shallow, irregular chest breaths, which strain the body and cause waste products to build up in the blood, which in turn makes us feel anxious.

You can demonstrate successful abdominal breathing to yourself by placing one hand on your stomach and the other on your chest. Now focus on your breathing. If you are breathing abdominally, the hand on your stomach will rise and fall as you inhale and exhale. But if you are chest-breathing, you will notice that the hand on your chest will move as you breathe.

Step 3.

Direct your attention from everyday thoughts by mentally repeating a neutral word—a word like "one," or "relax," or "peace," or "heavy"—with each outbreath. Assume a passive attitude. In other words, let relaxation occur at its own pace; don't "try" to relax, and don't worry about whether relaxation is occurring. If distracting thoughts occur, disregard them and return attention to the mental focusing device.

When you're ready to try it for yourself, you may want to tape-record the following, more detailed instructions—read them aloud very slowly—and play them back when it's time to relax:

Sit in a chair or lie on a bed or floor and get comfortable. Close your eyes and direct your attention to the toes and feet. Begin to imagine a wave of relaxation moving through them. You may notice a sensation of warmth, tingling, or heaviness there, or you may simply feel your feet against your shoes or the floor.

Now, visualize this wave of relaxation moving up through your calves, and then up to your thighs. Take time to appreciate how warm and good it feels. Let it come at its own pace. Now, feel it move up to your stomach, your chest, and then over to your back. Your upper body is becoming more and more relaxed.

Feel the relaxation moving out to the hands, where once again you may notice warmth, tingling, or heaviness. Perhaps you feel your hands resting against your body or your chair or bed. Visualize the relaxation spreading up to your forearms, upper arms, and shoulders. Take a moment to truly experience the good feeling of the relaxation. Next, let the relaxation flow to your neck and jaw, your cheeks, then your eyes and forehead.

Relaxation has now streamed throughout the entire body. Every inch of you is becoming more and more relaxed. Again, take a moment to feel how truly relaxed you are.

Focus your attention now to breathing. Note that it is becoming more rhythmic. As you inhale, feel your stomach expand; as you exhale, it contracts. Inhale, it expands; exhale, it contracts. Take a few moments to focus your attention on your abdominal breathing. If your mind drifts back to everyday thoughts, simply let the thoughts go and return your attention to your breathing.

Visualize a balloon in your abdomen. As you inhale, the balloon fills with air; when you exhale, it gets smaller. Stay focused on your abdominal breathing.

By now you are breathing even more rhythmically. It may help to pick a word to repeat silently to yourself each time you exhale, a word such as "one" or "relax." This word will serve as a focus of attention to direct your mind away from any other thoughts. Continue to focus on your breathing and the word.

Whenever it feels right, take a slow, deep breath, and then slowly open your eyes.

What was your experience like? When people first elicit the RR, they usually report a sense of physical relaxation, as their muscles untense and their breathing slows. During deep relaxation, you may feel that you are neither really awake nor really asleep. You may lose awareness of surroundings, thoughts, or the mental focusing device, and enter a state similar to stage 1 sleep.

Your mind may wander at first. That's normal. With practice, though, your ability to focus your attention will improve, and your thoughts will slow and begin to drift pleasantly.

At first, the feelings of relaxation you derive from the RR may be only temporary. Some of the beneficial effects of the RR are immediate and obvious; others occur only over longer periods of time and are more subtle. Friends may begin to notice that your fuse is not as short. Within as little as a few weeks, the body's stress hormones become less reactive. Your unconscious emotional activation is reduced, and the calming effects of the RR begin to "carry over" and extend throughout the day. As a result, you may begin to observe:

• An increased awareness of tension when it does occur, decreased stress reactions, and a greater ability to counter the tension by summoning relaxation quickly and automatically.

• Less anxiety, anger, frustration, and stress-related symptoms, leading to a generally improved daytime mood.

• An increased sense of control over stress, which leads to greater confidence in your ability to control your mind, body, and emotional reactions.

Guidelines for Establishing Daytime RR Practice

The more consistently you practice the RR, the greater the benefits to your sleep, your health, and your daily life. If you miss a day or two a week, don't become concerned, but remember: Practicing just a few times per week is usually not sufficient to counter the daily effects of stress on the mind and body.

The most difficult part of establishing regular RR practice is finding the time to do so. The tyranny of the Thinking Mind tells you that the day is too busy for relaxation, and will try to make you feel guilty or "unproductive." Think of the RR as something that will improve not just your mood and your health but your performance as well. It is something that you need and deserve. If you still can't find time for the RR, you are probably one of those who need it most.

Here are some additional guidelines to help you realize the many potential benefits of the RR:

• Allot ten to twenty minutes per day for the RR. Most people simply can't relax and quiet the mind in less time. As you gain more experience, you'll reach a greater degree of relaxation more quickly.

• Practice the RR in a comfortable position and quiet place where you are least likely to be disturbed by noises, telephones, children, or pets. Most people practice the RR at home, but it can be practiced in a library, conference room, or office. If you practice in the same place regularly, you'll begin to associate the RR with that place and you'll more easily make it a habit.

• Experiment to find the time of the day that works best, then designate that time as your regular RR period. By practicing at the same time each day, you'll make the RR part of your habitual routine. Some find it helpful to start the day with the RR; others discover that practicing later in the day, when stress responses have accumulated, is more effective. Afternoon RR practice may also satisfy the biological need for a mid-afternoon counterbalance to afternoon stress.

• Additional RR scripts are included in Appendix C; experiment until you come upon one you like. After more than twenty years of teaching patients the RR, I have found that the most effective way to learn it is by using a cassette tape, which makes it easier to focus attention and keep the mind from wandering. Consider making tapes of the RR scripts in Appendix C or purchase an RR cassette tape at a bookstore. The RR tape that I recorded and use with my own patients, and that I have used in my research studies, can be ordered with the form in the back of this book.

Minis: Eliciting the RR Anytime, Anywhere

Finally, there's actually a way to elicit the relaxation response in situations where you only have a few seconds available. We call it a "mini," and it allows you to significantly reduce your stress at a stoplight or in a traffic jam; during an argument; while waiting in line; just before giving a speech; or as you walk into a crowded room.

A mini involves taking a few moments to relax the muscles in the body (particularly the neck, shoulders, and face), then practicing abdominal breathing and mental focusing techniques. You can do this sitting or standing, with your eyes open or closed.

Take a moment now to try it.

Begin by closing your eyes and visualize your legs, arms, upper body, and face, relaxing. Next, focus on abdominal breathing for a minute or two. With each out-breath, you may want to mentally focus on a word. If the mind wanders to distracting thoughts, let them go and return attention to the breathing and word. Now open your eyes —that's all there is to a mini.

Although minis may not be as effective as the RR in terms of producing deep relaxation, they offer two advantages. First, they can be used anytime and anywhere to cope with stressful situations as they occur. Second, minis can be used more frequently than the RR and may therefore be more effective in countering frequent stress responses than practicing the RR only once a day.

Practicing a mini is easy; *remembering* to practice one is more difficult. There are numerous occasions when we simply don't recognize tension and forget we have this tool available. Therefore, it is sometimes necessary to use cues to remind ourselves of it.

Place a small piece of colored tape on the crystal of your watch. Each time you glance at the watch, the tape will act as a cue to practice a mini. Other cues you might use are stoplights or situations in which you find yourself waiting (television commercials, lines, on hold on the telephone, waiting rooms). A sticker on the telephone or reminders taped to the refrigerator or bathroom mirror are also good signals; by using them, you can make minis become an automatic habit whenever you experience tension.

Minis can improve your awareness of subtle levels of tension, distract attention from stressful thoughts, and keep your attention focused in the present. And, they make summoning the RR more a automatic process. So practice them regularly.

Reanimating an Ancestral State of Mind

The relaxation response is the most direct route to gaining access to your Ancestral Mind. It quiets and clears the mind and improves the ability to focus and concentrate, so your performance and problem-solving skills may improve. As you slip into your calmer Ancestral Mind, you'll find a sense of clarity, stillness, and inner peace. Over time, this heightened sense of calm can begin to extend throughout the day and serve as a foundation for a more positive outlook on life. By turning off the internal monologue, you may find yourself discovering a new sense of self, one that is often described as stronger, more connected, and somehow "higher." You may begin to feel less self-absorbed. You'll be better able to perceive similarities, and the fundamental unity, between yourself and others. Ultimately, you may feel more unified, harmonious, integrated, and whole. This change in self-awareness can foster greater self-esteem and self-acceptance, inner strength, and a healthier outlook on life. Thus, the RR can be a tool not just for relaxation but for personal transformation and growth.

As the RR quiets your internal monologue, you may also experience a shift in awareness of the external world, becoming aware of things that are normally registered unconsciously. You may, for example, be more likely to hear a deeper meaning in what other people are saying, and become more attuned to the body language of others and their nonverbal communication. In this newly relaxed state, you become more fully absorbed in the here and now, which is more peaceful than worrying about the past and the future. You experience the physical world more directly and vividly. Everyday moments take on a more immediate sense of joy, vibrancy, and meaning.

Try the relaxation response and see for yourself. It is a vital prescription for resting and relaxing, becoming absorbed and entranced, and hearkening to the wisdom of the AM.

Chapter Ten

Before Words, Images Were

During my postdoctoral fellowship at Harvard Medical School, I accompanied a team of scientists on a research expedition to Sikkim, a small Indian state near Bhutan and Nepal, to study a group of monks who had been filmed exerting an uncanny degree of control over their bodily functions. On a day that measured 40 degrees Fahrenheit, dressed only in a loincloth, each monk wrapped himself in a sheet soaked in ice water. Then, using nothing but their meditative powers, the monks proceeded to raise their body temperatures enough to steam the sheets dry.

Using instruments we'd brought with us from Boston, we determined that the monks were able to alter their brain wave patterns and lower their oxygen consumption significantly during this form of meditation, which they called *g Tum-mo* yoga, or "fierce women" yoga.[1] They achieved this by first eliciting the relaxation response, and then visualizing what they call *prana*, meaning wind or air, entering the "central channel" of the body. It was to confront and calm these winds that they ignited "internal heat," which they saw as burning away defilements and improper thinking.

Halfway around the world, we use similar techniques of visualization to help patients deal with any number of medical problems.

Not long ago a woman I will call Marilyn came to us for treatment of Raynaud's disease, a conditon characterized by painful episodes of cold hands caused by constriction of blood vessels, especially during cold weather. I began working with Marilyn by administering biofeedback treatment, in which she was connected to a thermal biofeedback machine that measured her hand temperature and displayed it as a vertical bar on a computer monitor. When Marilyn's hands cooled, the bar fell; when her hands warmed, the bar rose.

Over a series of sessions, Marilyn learned that, when she thought about mental images of warmth, her hands became warmer. If she tried too hard, however, her hands became colder. In time she mastered the proper balance, and soon she was warming her hands from 70 to 95 degrees in as little as ten minutes. With further practice, she was able to warm her hands without the biofeedback machine, and could soon abort a Raynaud's attack using nothing but her mind's ability to direct visual imagery.

Just as the Ancestral Mind speaks to us in the form of imagery, so, too, can we learn to use imagery to speak to the Ancestral Mind. In this way we can not only evoke the relaxation response, but also open a direct channel to unconscious processes and exert far greater control over our emotions and our mind/body interactions.

■ Language Is the Voice of the TM, Imagery the Voice of the AM

Visual imagery is a form of symbolic mental processing that does not depend on the use of language. We visualize everything from goals to activities, desires, and wishes in our mind's eye. When we dream, daydream, remember, and plan, we use imagery. Imagery is a means of storing and expressing certain types of information and mentally rehearsing events in our lives.

Imagery is not unique to the Ancestral Mind. The Thinking Mind uses imagery to remember past events and foresee future ones, solve problems, and learn new skills. When a scientist visualizes the structure of a molecule, a tennis player rehearses her serve, or you think about what to wear to a party, visual imagery is being employed. Our technological advancement has always depended upon our ability to visualize and imagine a desired outcome. But the Thinking Mind's primary language remains words, while the Ancestral Mind was using visual imagery long before language evolved.

The Ancestral Mind uses preverbal imagery to convey emotions, feelings, and intuitions. As we explored in chapter two, the amygdala responds to such symbolic information as it is conveyed from the thalamus. There are even those who believe that symbolic, preverbal data is stored in the Ancestral Mind as a form of genetic memory. In his book *Genes, Mind and Culture*,

Harvard evolutionary biologist E. O. Wilson argued that preferences for certain patterns or colors, for example, have been conveyed genetically. He calls these small units, the cultural equivalent of genes, "memes": fundamental human symbols and images that have been held in common throughout our evolutionary development, and that are a form of "species knowledge," passed from generation to generation in the same way that a bird species' distinctive song is transmitted. At a less technical level, preverbal imagery is also the basis for the Jungian archetypes we discussed earlier, the stuff of myths, spiritual visions, and the deeper human realities that seem universal and eternal.

We humans have evolved a complex faculty for using imagery in our emotional lives. Virtually everyone has some awareness of how images in the mind can affect thoughts, emotions, and the body. You can demonstrate the powerful effect of imagery with this exercise:

> Close your eyes and imagine holding a bright yellow lemon. Feel the texture of the fruit in your hand. Imagine cutting it with a knife. Notice its color and smell. Now visualize yourself biting into the lemon, and tasting the pulp. The tartness of the juice is so intense that you begin to salivate.

Mental images in our mind's eye can elicit even more powerful physical reactions. Erotic images conjured can cause sexual arousal. Visualizing a future event can cause us to become excited or anxious. Recalling a past event can evoke pleasure, embarrassment, or, in the case of a traumatic event, fear and emotional anguish.

Visual imagery is a natural endowment whose full powers we lose too often and early in life as a consequence of our emphasis on the TM's verbal processes. Words have become our primary way of knowing and understanding the world, leaving imagery to the unconscious. The excessive stimulation that the TM forces on us—both from the modern world it has created, and from its own internal chatter—also makes it harder to consciously hold an image in our mind.

Visualization may be the most ancient healing technique used by ancestral man. In traditional cultures, shamans, spiritual leaders believed to possess special skills that enable them to heal the ills of an individual or a community, use techniques for entering into a trance in order to obtain unique knowledge of an illness or problem. This kind of healing trance is a nonrational, intuitive state that involves extraordinary problem-solving abilities, and can be entered and exited at will. Like the reveries and other forms of deep relaxation we've already discussed, this state is "dreamlike," somewhere between wakefulness and sleep, and rich with vivid imagery. The shaman uses the "journey" into this altered state of consciousness to receive and interpret unconscious images and symbols both to diagnose the sources of illness and to aid in visualizing a cure.

Just as repetitive mental stimulation is used to induce the relaxation response, the shaman relies upon ritualistic chanting or the beating of drums as a way of entering this altered state. His perceptions of time and space are altered; he experiences a loss of his everyday self and he attains a "higher" state of awareness, which is similar to peak experiences, mystical experiences, and all the other ancestral states of mind we've discussed.

■ The Mind/Body/Image Connection

Studies conducted as early as the 1930s demonstrated that imagery could alter heart rate and respiration.[2] More than fifty years ago, Dr. Edmund Jacobsen, the developer of a relaxation technique called "progressive relaxation," established that merely visualizing an action can make the appropriate muscle respond.[3] Other studies of imagery showed that while reading frightening scripts to subjects did not affect them physiologically, asking the same subjects to visualize the frightening content caused their heart rate and skin sweating to increase significantly.[4]

In research conducted at Yale University, Dr. Gary Schwartz and his colleagues found that pleasant images create a response in the muscles of the cheek that are responsible for smiles, whereas unpleasant images produced a response in the muscles involved in frowning.[5]

Other studies have shown that when people who are phobic about spiders imagine scenes involving these creatures, heart rate and skin sweating increase dramatically.[6] This is similar to the reaction known as post-traumatic stress disorder (PTSD), in which nightmares and flashbacks of a traumatic event cause the individual to reexperience the physical effects associated with it. Brain imaging studies have determined that PTSD patients show emotional activation in the mammalian brain in response to these disturbing images, which explains why imagery-based treatments can help in managing their symptoms.[7]

The most extensive evidence on the effects of imagery on emotions and physiology concerns its use in a technique for overcoming fears called systematic desensitization. This method, which was developed by Dr. Joseph Wolpe and has been validated in hundreds of research studies,[8] is based on two concepts: first, that anxiety and phobias are associated with physiological tension and fearful imagery; and second, that it is impossible to feel fear while one is mentally or physically relaxed.

In order to diminish a phobia about snakes, for example, patients create a hierarchy of images ranging from low to high anxiety: seeing a snake, approaching a snake, then holding a snake, and so on. By pairing the relaxation response with the fearful images in a gradual fashion, patients learn to visualize the fearful image without the associated anxiety. Thus, they "unlearn" the association between the image and anxiety and, instead, learn to associate emotional and physiological relaxation with the image.

A patient of ours named Paul had a fear of needles. Just the thought of a needle could make him sweat, and simply waiting in the doctor's office to receive an immunization had caused him to faint. We taught Paul the relaxation response, then taught him to pair a sequence of images with the RR: going to the doctor's office, sitting in the waiting room, being called into the exam room, seeing the needle, and so on.

After several weeks, Paul was able to visualize himself receiving an immunization without feeling anxious. He later learned how to use mini-relaxations and positive imagery to keep himself calm when he actually went to the doctor's office. With this preparation, he was able to receive an injection without fainting for the first time in years.

Researchers at the University of California at San Francisco taught patients to use relaxation and visualization techniques aimed at reducing the breathing problems associated with asthma. One group of patients was taught to visualize images of breathing easily through unobstructed bronchial tubes. Another group was taught to imagine the cells that trigger asthma attacks becoming calm and stable. Both groups experienced improvement in their symptoms.[9]

In a study of the pain associated with receiving bone marrow transplants for cancer, patients who practiced relaxation imagery reported less pain than patients who received standard care.[10] There is also some evidence that imagery may affect immune functioning. Cancer patients whose disease had spread were taught to practice a relaxation and guided imagery exercise over a one-year period while monthly blood samples were collected and analyzed. These patients showed increased immune response as measured by increased natural killer cell activity.[11]

Because scientists are well aware of the fundamental link between imagery and emotional response, they routinely instruct subjects to visualize both pleasant and unpleasant events in order to study the associated neuroanatomical correlates of emotions. These studies consistently reveal that the mere act of holding an emotionally charged visual image in the mind's eye is sufficient to produce energy consumption in the brain, as well as direct, measurable changes in the reptilian brain, the mammalian brain, and the cerebral cortex.[12] In a similar vein, research by cognitive psychologist Dr. Stephen Kosslyn at Harvard University demonstrates that visualizing an object activates the same brain areas that are activated when the object is actually seen.[13]

We know that imagery affects brain activity, heart rate, blood pressure, respiratory patterns, oxygen consumption, carbon dioxide elimination, skin sweating, blood flow, activity in the gastrointestinal tract, sexual arousal, and immune system functioning. It can also affect our thinking. While imagery represents a means of establishing a dialogue between conscious and unconscious, between mind and body, it takes a bit of practice to become proficient at using it.

■ Using Imagery to Evoke the RR, Influence Emotions, and Converse with the AM

Guided imagery is an excellent way of evoking the relaxation response. With this method, you deliberately summon a particular image in your mind that serves as a mental focusing device and diverts attention away from the TM's mental monologue. Some of us are very visually oriented and can easily picture clear, precise images in our mind's eye; others see only imprecise images or just "sense" them. Whatever you experience when you practice guided imagery is suitable for you; there is no right or wrong way to visualize.

The following script combines a visualization of an idyllic natural setting with breathing and muscle relaxation techniques to elicit the relaxation response. This is another exercise that might be most useful for you to tape-record for repeated usage.

Close your eyes and direct your attention to your toes and feet. Imagine a wave of relaxation beginning to move through your feet.

Now the relaxation spreads through your calves and thighs, then to your stomach, chest, and back. Note how tension is disappearing. Sense how relaxation is flowing now into your hands, forearms, upper arms, and shoulders. Now the relaxation moves to your neck, jaw, cheeks, eyes, and forehead. Tension is vanishing from your entire body.

Concentrate now on your breathing, noting that it is becoming more rhythmic and abdominal. Also notice that, as you inhale, your stomach expands; as you exhale, it contracts. Inhale, it expands; exhale, it contracts. Take a few moments to focus on your abdominal breathing.

If your mind drifts to distracting thoughts, let the thoughts go and return your attention to your breathing. It may help to repeat a word silently to yourself each time you exhale, such as the word "one" or "relax." This word will serve as a mental focus. Take a few moments to concentrate on your breathing and on the word you have chosen.

Now that you are relaxing more deeply, imagine yourself in a tranquil place in nature, one that you have visited or would like to visit: a meadow, a beach, a mountain. It can be real or imaginary; it need only be somewhere you feel completely comfortable and peaceful. Take a few moments to look around and consider the beauty: the sky, trees, animals, flowers, or whatever you observe. Are there any sounds, such as water, wind, or birds? Are there any smells, such as flowers or grass? Listen to your surroundings, and sense the warmth of the sun against your skin. Take a few moments to be at peace in this soothing environment. You are safe and comfortable here.

Now sit or lie down in this serene place. You are becoming more deeply relaxed. You feel calm and peaceful. Enjoy this safe and wonderful refuge, to which you can return anytime you need to relax.

Take one more look around, then draw a slow, deep breath, and slowly open your eyes.

What was your experience like? Were you able to picture a relaxing scene? Did your mind wander to other thoughts? Did you sense mental or physical relaxation? With practice and exploration of different images, you will be able to relax more consistently using guided imagery. You may want to incorporate a particular nature scene into your regular practice of the relaxation response, or you may want to experiment with a variety of scenes.

Guided imagery can also be used when you practice mini-relaxations or feel tense during the day. Many of my patients find that this technique is not only a pleasant respite from daily tension; it also allows them a greater sense of control over stress, which is a powerful stress-reducer itself. These respites likewise heighten awareness of tension and relaxation and stop mental chatter.

Guided imagery is an effective method of managing situations in our lives that cause anxiety. Employing it for this purpose is similar to systematic desensitization, in which patients pair relaxation with fear-producing imagery in order to alleviate the anxiety associated with the image. Let's say you're preparing to give a talk or interview for a job. The anticipatory anxiety you may experience in the weeks or days preceding these events can be exacerbated by negative, stressful imagery held in the mind's eye, which causes us to worry about events that may never actually occur and to trigger stress responses.

Think about an event that causes you to feel anxious. Close your eyes and examine your mental image of that event. Do you picture it turning out badly? Does the image contain elements of fear? By becoming more aware of such images and consciously altering them to appear more positive, you can diminish anxiety. It's simply a matter of mentally rehearsing these events in the most favorable light. Many athletes do precisely this in preparing for competition. They visualize themselves flying over the pole-vaulting bar, or they picture their foot making contact with the football and sending it flying directly between the goalposts—or whatever it is they most want to achieve. This kind of imagery heightens the sense of self-control and mastery.

You can practice guided imagery in anticipation of anxiety-producing events whenever you elicit the RR. Just before opening your eyes at the end of the session, picture the event in your mind's eye and become more aware of precisely how you are imagining that it will take place. Then, consciously change the image so that it has a more positive outcome and hold that image in your mind.

Suppose you are anxious about giving a speech. After you elicit the RR, examine the image you have been holding of the talk, and then alter that image so that it has a more positive outcome. Visualize yourself, for instance, walking up to the podium, looking relaxed and confident, and delivering the speech slowly and clearly; see the audience looking interested and applauding enthusiastically at the end of the talk. You can use the same procedure for a job interview or a showdown with your boss, or any other situation that is causing you concern. Positive, confident imagery is a powerful tool for reducing anticipatory nervousness and helping you to cope effectively with, and think optimistically about, the event at hand.

A final type of visual imagery, called spontaneous imagery, involves becoming more conscious of the free flow of images that occurs during the relaxation response. Whereas guided imagery is a matter of "talking" to the AM by consciously creating images in the mind's eye, spontaneous imagery requires being receptive and "listening" to the AM.

When you elicit the relaxation response, you may find yourself drifting between reverie (theta EEG) and a more alert but relaxed state (alpha EEG). Pay attention to the images that come to mind when you drift out of reverie. Although these hypnagogic images may be hard to "catch," they can help you to visualize a new idea or solution to a problem, or to gain intuitive insight into emotions and feelings. Spontaneous imagery is how many people view the act of prayer: It's not so much talking to a higher power as it is listening to that power.

The other useful time to observe spontaneous imagery is just before you open your eyes after eliciting the relaxation response. At that point you are likely to be in an alert but relaxed (alpha EEG) state in which images flow freely through your mind. Observe these images, too, for they may also represent unconscious emotional information or valuable insights.

As you become more conscious of how your mind uses imagery, you will become increasingly familiar with the language of the AM. You will notice that imagery accompanies much of your thinking—even when your eyes are open and your attention is directed to the external world.

With a greater awareness of the content of your personal imagery, you can use these techniques to modify your behavior, your emotions and thoughts, and your physiology. As the forgotten language of the Ancestral Mind, imagery touches the deepest part of ourselves. It is an untapped resource that can enable us to gain new connections to and perspectives on our inner world, and lay claim to a deeper, more basic form of experience.

Chapter Eleven

The Ancestral Mind's Minimal
Daily Requirements

There are three natural "nutrients" that were a vital part of ancestral life, but that have become marginalized in our modern world: music, light, and exercise. Almost anyone today would agree that these are pleasant and even valuable things. The point I want to make here, however, is that they are as essential to our health as the full range of essential vitamins. We can't just settle for lip service in appreciating them; we need to bring them actively back into our lives as part of the Ancestral Mind's prescription for health and well-being.

In this chapter we're going to look at each one in turn.

■ Music: Our First Language

Just as we are born with the capacity to process language, we are born with the ability to apprehend music. Our brain unconsciously translates patterns of sound energy into the basic musical components of melody, harmony, and rhythm. Like spoken language, music is a sound-based form of communication involving a set of rules for combining elements in an infinite number of ways.

At four months of age, infants exhibit musical preference, suggesting that we enter the world innately prepared to respond to rhythm.[1] Harvard University psychologist Howard Gardner, author of *Frames of Mind: The Theory of Multiple Intelligences*, contends that musical creativity represents a basic form of intelligence that is just as important as linguistic and mathematical intelligence.[2]

Although there is no music "center" per se in the brain, music perception involves various areas in the cerebral cortex, the mammalian brain, as well as in the most ancient reptilian brain. But it is music's effects on the mammalian brain, in particular, that are responsible for its profound ability to evoke an emotional response.[3]

To understand just how deep and primal the roots of music are in us, consider that the existence of music predates the emergence of modern humans by millions of years, for it was part of the natural environment in which we evolved. Whales and birds produce songs that are remarkably similar to our own: rhythmic utterances that are constructed according to rules not unlike those followed by human composers. Whales have a vocal range of seven octaves, and they use musical intervals between notes that are comparable to the musical scales that we use. Because whales create songs that are about the same duration as a ballad or one movement in a symphony, it may be that these marine mammals have an attention span similar to ours. Whale songs even contain repeating refrains, suggesting that they use devices similar to human storytellers to help them remember complex material. Because whales from different oceans sing different songs, scientists can actually identify the year and the ocean from which a song came by comparing it with a collection of previously recorded whale songs.[4]

Bird songs also show remarkable parallels with human music; they use the same rhythmic variations and combinations of notes. Birds also recognize and memorize musical patterns, pass songs from one generation to another, and share their songs, just as humans do. Some birds can even play musical instruments. The male palm cockatoo of northern Australia breaks a twig from a tree, then shapes it into a drumstick. The bird then selects a hollow log and, holding the stick with its foot, drums on the log as part of its courtship ritual.[5]

One of Mozart's musical companions was a starling, a bird known for imitating and creating songs. While Mozart was composing a piano concerto in G major, the starling was learning parts of it. In one of his notebooks, Mozart recorded the starling's interpretation of the final movement of the piece. Although many of the starling's notes were the same as the composer's, the bird also improvised, and quite "beautifully," as Mozart noted.[6]

Music is part of our biological makeup, encoded in our genes as part of what makes us human, and as part of what unites us with other species. As one of the oldest and most universal forms of human experience, music has been used throughout our history to affect emotions, physiology, and states of consciousness. Music can pick us up when we are down and calm us when we are anxious. It can inspire romantic or spiritual feelings and can move us from tears to joy.

Music brings communities together, and has played a role in the celebration of harvests, births, and marriages. It has been used to heal and to invoke the spiritual world, as well as to mourn. Ancient cultures such as those of the Sumerians and Babylonians believed that music was the master metaphor for life and creativity; some of the early philosophers saw music as a primal expression of cosmic creation. In his dialogue *Timaeus*, Plato tells how the Creator made the world according to musical properties, setting in motion "the music of the spheres." In ancient times, the most dismissive assessment of a man was that "he hears no music."

Recent scientific evidence suggests we may have played musical instruments before we spoke. Cro-Magnon and Neanderthal man made flutes out of animal bone 53,000 years ago; these instruments were much more advanced than any other form of technology of the time, including spear points and carving tools. Ancestral man obviously considered music a central part of existence, a pursuit worth devoting time and attention to. Some scientists speculate that the ability to memorize and recognize musical patterns may have been the foundation for oral traditions, passed from generation to generation like birdsong, well before the development of language.[7]

Imagine our distant ancestors sitting by a fire under a luminous starfilled sky, listening to the mesmerizing melody of a flute that imitated the sounds of birds. This was a time when music connected ancestral humans to one another, integrating the rhythms coursing through their minds with all the pulsing rhythms of nature. That was a time of genuine balance, a kind of inner and outer harmony that we would do well to restore.

Scientific studies have shown that music is highly effective in undoing a bad mood, raising energy levels, and reducing tension.[8] When used in the operating room, music increases relaxation in both surgeons and patients, resulting in fewer surgical complications. It also lessens the need for pain medication, and speeds recovery.[9] During childbirth and dental procedures, music reduces pain and anxiety.[10] In another study, soothing music was played during surgery in the operating room, reducing by half the amount of sedative needed by patients.[11] One researcher estimated that the calming effect of music was equivalent to an intravenous dose of 2.5 milligrams of Valium.[12] Another study found that music that was introduced before bedtime reduced the need for sleeping pills in geriatric residents by two thirds.[13]

A review of the effects of music in the journal *Science* concluded that music can positively affect test performance, blood pressure, pain perception, and heart rate.[14] In a study conducted by Dr. Avram Goldstein at Stanford University, approximately one-half of subjects reported a sense of euphoria and "thrills" in response to music, which they characterized as a chill, shudder, or tingling sensation associated with sudden changes in mood or emotion. In some cases, the feelings were described as "awe." Goldstein also surveyed 250 people and found that such thrills occurred more in response to music than in response to any other stimuli, including movie scenes, sexual activity, or beauty in art.[15] All of this data reinforces our belief that the neural substrates of music lie deep in the evolutionary history of the Ancestral Mind, and that music is crucial to our emotional health.

Music may exert its therapeutic effects in a number of ways. Many of us have had the experience of a piece of music triggering a feeling or memory of an emotionally meaningful event (such as a wedding or funeral) or a romance (when that was "our song"). Cognitively, many kinds of music can evoke the relaxation response by absorbing our attention and directing our thoughts away from worries and concerns. Physiologically, music can act as a repetitive mental stimulus that provides constant input into the sensory receptors, reticular formation, and thalamus, all of which induce relaxation at the mammalian and cortical levels. By quieting the brain, music produces muscular and physiological relaxation.

The emotional and physiological effects of music are, of course, dependent in part on the type of music you listen to. People generally describe high-pitched or major-key music as happy and low-pitched or minor-key music as sad. While faster, louder music increases heart rate, blood pressure, and alertness, slower, quieter music is more likely to induce a sense of calm or deep physical relaxation. Although music can induce positive emotions ranging from excitement to serenity, calming music can help to evoke the relaxation response. Although the latter effects can alleviate stress-related physical symptoms such as pain or muscular tension, any kind of music that elicits positive emotions will enhance emotional well-being by counteracting the effects of unhealthy negative emotions.

When I used music as a comparison condition to study how the relaxation response affects brain waves, it did not produce the same magnitude of changes in EEG activity as the relaxation response, but it did yield the same general *kinds* of changes—namely, increases in theta EEG power.[16] You may recall that theta EEG is associated with deep relaxation, hypnagogic and primary process states, and the transition from wakefulness to sleep (stage 1 sleep), all of which are synonymous with conservation of energy in the brain. Music likely cultivates mental states of quietude by eliciting the RR. (Interestingly, some individuals seemed to do better with music in attaining calmness than with more traditional RR techniques such as muscle relaxation, breathing, and mental focusing, as validated by increases in their theta EEG power.)

The Magic of Melody in Daily Life

Whether we hear it on the radio, in the jingles that are part of commercials, the background themes to television shows, or piped into stores, we spend some part of our day listening to music. In some instances, we consciously attempt to regulate our moods by choosing the music we hear; at other times, the music is chosen for us, which may not have positive effects on our behavior and well-being, particularly if it is a type of music that we don't like.

If you are spending some time each day listening to music that you enjoy, you are already reaping some of its beneficial influence. But there are additional steps you can take to more consciously and deliberately employ music to help you connect with your Ancestral Mind.

Try playing music in the background when you practice muscle relaxation, breathing techniques, and the mental repetition of your focusing word or phrase. Many commercially available relaxation tapes offer music especially selected for this purpose. You may discover that music works as well as or better than other mental focusing devices in eliciting the RR, or you may want to give yourself more variety by alternating between music and your special focusing word or phrase.

One of the best times of the day to nourish the Ancestral Mind with music is before bedtime. Our brain needs a winding-down period before bedtime to turn off the TM and allow sleep to gradually replace wakefulness. Music can not only reduce tension from the day but also improve our ability to fall asleep and sleep soundly during the night.

Many people, of course, make use of music intuitively. My purpose in calling attention to it as a pathway to our emotional world is to reinforce the practice, to assure you that its benefits are not inconsequential. Music can be seriously therapeutic, and you should make the most of those benefits by making music a regular part of your daily routine.

■ Let There Be Light

Natural light is the normal environment for the Ancestral Mind. During the entire course of our evolution as hunter-gatherers, we were exposed to cycles of sunlight during the day and darkness at night. Because natural light energizes and sustains us, our mood on sunny days tends to be more upbeat than on dull, gray ones. For the same reasons we often experience the irresistible urge to get away to a sunny climate in the dead of winter.

With the advent of modern technology, however, we get little exposure to natural sunlight, as we spend our days working or living indoors. Studies have shown that, no matter where people live, they obtain only one hour of sunlight on average during the day. Meanwhile, the "light pollution" that is typical of most urban environments means that many people are also not exposed to true darkness anymore. At night we live with indoor light and the constant illumination of city-lit skies. When was the last time you saw the luminous Milky Way or the brilliance of the stars, something that our ancestors witnessed regularly?

A brightly lit room has about 500 luxes of light (a lux is the equivalent of the light from one candle), compared to 10,000 luxes of light that are present at sunrise and 100,000 at noon on a summer day. To the brain structures of the Ancestral Mind, spending the day indoors is equivalent to spending the day in darkness. The problem is compounded in the late fall and winter months, when days grow shorter and many of us curtail outdoor activities. Our distant ancestors evolved on the savannas of Africa, but many of us now live in northern latitudes where ambient sunlight is significantly diminished in the winter.

The initial scientific studies on the effects of light on mood involved depression. In the early 1980s scientists at the National Institutes of Mental Health found that a patient who experienced depression annually during the fall and winter months (what is called "seasonal depression") improved after daily treatment with exposure to bright light.[17] That single case study has since led to the identification of a specific condition called *seasonal affective disorder*, or SAD, which is characterized by depression and sleep disturbance in the winter months. It also has led to extensive testing of the efficacy of bright light therapy, and there is now a scientific consensus that bright light therapy is effective for SAD.[18]

Studies show that for the general population, mood and energy, even among nondepressed individuals, are at their lowest during the short and dimly lit days of winter.[19] Because inhabitants of northern zones spend almost six months a year in wintry lighting conditions, SAD symptoms may be present a majority of the time. People surveyed in New York City rated the degree of seasonal change in the duration of their sleep, in their social activity, mood, weight, appetite, and energy by indicating when they felt best and worst, socialized the most and least, and so on. They also rated the degree to which seasonal changes presented a personal problem in their lives. Half the respondents reported lowered energy in the late fall through the winter, and a quarter of the respondents reported that poorer mood in the late fall and winter posed a personal problem.[20]

Some scientists believe that the lack of exposure to sunlight affects the population as a whole and may be contributing to an overall increase in mood disorders, possibly through disruptions in the balance of neurotransmitters such as serotonin. They also believe that widespread exposure to supplementary light, at least in the winter, might have widespread benefit.[21]

Despite the therapeutic effects of light on mood and sleep, we know surprisingly little about the underlying mechanisms by which bright light exerts its effects. Presumably, its antidepressant and mood-energizing properties are due in part to its action on an area of the brain called the suprachiasmatic nucleus, which is located in the hypothalamus. This nucleus contains a pacemaker that acts as a "clock for all seasons" by being tuned to seasonal variations in light.[22] For mammals, changes in light and darkness play an important role as a chemical signal for breeding and for other behaviors such as eating, sleeping, and putting on weight. In many animals, including humans, the pacemaker can detect seasonal changes in day length and make corresponding changes in the brain.[23]

Besides affecting our mood, light affects our sleep, which is important for restoring energy and optimal neural functioning. The daily cycles of light and darkness act on melatonin, a naturally occurring hormone found in the brain. When sunlight enters the eyes, melatonin levels decrease, which signals body temperature to rise and promotes wakefulness. Darkness causes melatonin levels to rise and body temperature to fall, which promotes sleep.[24] When we reduce our exposure to bright natural light and true darkness, we alter our melatonin secretion, thus altering body temperature rhythm, which can exacerbate sleep difficulties. This phenomenon explains why up to 90 percent of the blind experience sleep problems.

Aside from the most obvious way of increasing your exposure to natural bright light—spending more time outside—you may want to consider the use of artificial bright light boxes that can be rented from medical supply companies or purchased from an increasing number of manufacturers. Some insurance companies even reimburse the cost of these devices if they are prescribed by a physician for SAD or insomnia.

Bright light boxes contain special bulbs that emit 5,000 to 10,000 luxes of light, which is equivalent to the amount found naturally in a sunrise or sunset. They are used for about thirty minutes daily while reading or watching television to increase early- or late-day exposure to bright light. Several studies have demonstrated that using bright light boxes in the evening can delay the body temperature rhythm and effectively minimize early-morning insomnia. Other studies have demonstrated that the boxes are effective for SAD.

Bright light is easy to administer, free of side effects, and inexpensive. Once again, it is so simple a solution that we often don't take it seriously. But it is an essential part of the ancestral environment in which we evolved, and therefore an essential ingredient in our physical and emotional health.

■ Movement and the Mind

We are biologically designed for movement. For 99.9 percent of our evolution as hunter-gatherers, scavengers, and toolmakers, our survival depended on regular physical activity. Until only recently in our evolutionary history, hard labor was still a necessity as we learned to plant, sow, and harvest crops. Living close to nature was rigorous, and natural selection accordingly wove the need for that kind of rigor into our genetic makeup. That's why physical inactivity is so physiologically and psychologically unhealthful.

Nonetheless, physical activity has all but disappeared from daily existence in postindustrial society. The Thinking Mind's work environment is sedentary and indoors. Everywhere we look, some kind of electronic or otherwise mechanized equipment has taken over the physical work that humans used to do. The computers, automobiles, televisions, and other labor-saving devices that were intended to make life easier have instead made us fat and depressed. We work in climate-controlled offices, not open fields, and we fiddle with gadgets instead of engaging with other people, or with three-dimensional forms of recreation out in the real world. Natural selection did not refine our physical abilities for millions of years only to have us plop down on the couch. We don't even get up to change the channels on our televisions, or physically open our own garage doors. The convenience of elevators, escalators, drive-through everything, valet parking, and delivery services may be what we think we want, but it is not what we need.

Obesity—which was probably nonexistent in ancestral times—is today an epidemic. And obesity is more than a physical problem: It disrupts psychological functioning by depleting our energy and by lowering our self-esteem.

Despite the public awareness of the negative consequences of a sedentary lifestyle, more than half of adults get little or no exercise and at least 25 percent of adults (and an alarming number of children) are completely inactive and overweight.[25] In the past three decades, Americans have become even more overweight, even though surveys indicate that we consume fewer calories than in years past.[26]

In England one in five adults is now considered obese, which represents a tripling of obesity in twenty years. Two-thirds of English men and more than half of the women are clinically overweight. English officials estimate that 31,000 people a year are dying prematurely because of weight-related health complications, with poor diet and inactivity being the primary culprits. The average English citizen watches almost four hours of television a day, another rapidly increasing trend.[27]

In China, increasing wealth has come at the expense of health. Diabetes, high cholesterol, and heart disease are developing at record rates.[28] Although malnutrition levels are dropping, weight-related morbidity and mortality are rapidly increasing. The newly overweight and affluent Chinese have spawned a trend in which individuals spend one thousand dollars per week on resort-style health camps that return them to their ancestral roots through hoeing and harvesting in rice paddies.[29] Some Chinese have realized that, while the old life may have been demanding, its demands provided what was necessary for good health.

Many of our children grow up on fast food and sugar-saturated soda, under constant pressure from the food industry to eat candy, chips, sugary breakfast cereals, and other "foods" that were nonexistent in our diet for millions of years. Some children consume up to a thousand calories a day in soft drinks alone.[30] And as researchers have shown, obesity increases by 2 percent for every hour of weekly television viewing.[31] When you consider that children average almost thirty hours per week in front of their computers and in front of the tube (often while they eat junk food and view advertisements for more junk food), it's not surprising that 20 percent of children are now considered obese.[32]

Why are we gaining weight despite decreasing our caloric intake? Because we have dramatically decreased activity. Recent estimates suggest that our energy expenditure has fallen by three hundred calories per day in the last twenty-five years.[33] Dr. William Haskell, an exercise physiologist at Stanford University, has calculated that spending two minutes every hour at work e-mailing a colleague rather than walking down the hall to his office would accumulate eleven pounds of body fat every ten years.[34] Other researchers have calculated that using the television remote instead of getting up to switch channels can add a pound a year of fat; and, if we used just one hundred calories less per day while continuing to eat the same amount, we would gain about ten pounds per year.[35]

A significant percentage of Americans have chronic medical problems directly related to physical inactivity and obesity, including heart disease, high blood pressure, stroke, diabetes, osteoporosis, and some types of cancer; inactive people also have twice the rate of heart disease and heart attacks of physically active people. It has been estimated that a quarter of a million deaths each year in the United States, or one in eight deaths, is directly attributable to chronic diseases that are the result of a lack of physical activity. Many of these chronic diseases also markedly disturb quality of life by increasing the risk of depression.[36] It's not surprising that chronic conditions such as coronary heart disease were probably unknown in ancestral times and are still rare in contemporary hunter-gatherer societies.

Minding the Body

You don't have to start running marathons to experience a wide array of benefits from exercise. These include weight loss, improved physical appearance, improved general health, and a vastly improved outlook for longevity.

Regular exercise enhances health by improving cardiovascular functioning, bone density, and immune functioning. People who are physically active are much less likely to suffer from heart disease, hypertension, diabetes, osteoporosis, obesity, back problems, and colon cancer They are more likely to experience a stronger sense of well-being and exhibit better health-related factors such as body composition, glucose tolerance, and insulin sensitivity. Physical activity also increases high-density lipoprotein cholesterol ("good" cholesterol) and reduces triglyceride (the chief constituent of fats) levels and blood pressure.[37]

Becoming more fit can take twenty years off a person's chronological age.[38] Physically fit older adults enjoy better sex lives and have stronger immune systems. They are less likely to exhibit insulin resistance, which is a reduction in the body's ability to metabolize glucose and a risk factor for diabetes. Older adults who exercise are also less likely to develop obesity-related disease, and exhibit better memory, strength, range of motion, endurance, and lower blood pressure.

Mending the Mind

From an evolutionary perspective, physical activity not only gave us greater access to food and shelter; it also affected consciousness and motivation. Science has demonstrated that the positive psychological effects that accrue from exercise may be the result of alteration of brain neurotransmitters such as norepinephrine or serotonin, both of which play an important role in regulation of mood, arousal, and attention.[39]

With the greater mental energy and positive mood that result from physical activity, we were more likely to explore new territories, persist at difficult tasks, and engage in prosocial behaviors. It is very likely that this mood effect of exercise was itself an adaptation in the Ancestral Mind to ensure that we stay active. Once again, it is the Thinking Mind that has disrupted this natural process.

Exercise is an outlet for the body's excess tension, providing a healthy way to release anger and anxiety. It improves energy and has a tranquilizing effect that reduces anxiety more effectively than many medications designed to do so. Studies have found that the tranquilizing effect follows within five to ten minutes of completing exercise and lasts for at least four hours.[40]

People who exercise typically have a more positive body image and feel better about themselves. Psychologist Robert Thayer found that exercise is the single best method for getting out of a bad mood. He conducted extensive research on the effects of exercise on mood and found that just a five- or ten-minute walk has a very reliable and immediate mood-enhancing effect that persists for several hours.[41] The effect is the result of both a lowering of tension and an increase in energy, although the boost in energy is more consistent. (Although intense exercise reduces tension, it initially produces fatigue that is followed by a rise in energy.)

Exercise is even an effective treatment for people suffering from depression. In one study, mildly to moderately depressed people reported feeling better within one week of beginning an exercise program; they also showed greater improvement over time than mildly to moderately depressed people who received either short-term or long-term psychotherapy.[42] In a study that compared a popular antidepressant, Zoloft, to exercise for the treatment of mild to moderate depression, researchers found that exercise worked as well as the drug and did a better job of keeping the depression away once it had lifted.[43]

For most of us, getting out of our heads and accomplishing something physical brings everything together in a rhythmic merging of concentration, sensory awareness, and adrenaline. No distinction exists between mind and body; we feel focused and at one with our surroundings. In these instances, exertion can produce the highly desirable altered state of consciousness we've discussed previously as "flow." In this ancestral state of mind, the customary sense of time ceases, awareness merges with actions, and self-consciousness evaporates.

Many of us have experienced the true exhilaration and almost spiritual uplift that can come from physical activity. John Updike's novel *Rabbit, Run* features a scene in which the troubled protagonist, Rabbit Angstrom, is playing golf with a minister who's trying to help him. Rabbit hits a beautiful golf shot and cries out, "That's it!" almost as if he'd seen God. This young man can't seem to do anything right in terms of career or personal relationships, but give him a physical challenge, and he is in his element.

Emerging animal and human research suggests that exercise may also improve learning by improving blood flow, spurring levels of brain cell growth hormone, and stimulating the development of new neurons in areas of the brain involved in memory.[44]

Exercise as a Sleep Aid

Exercise also improves sleep by producing a significant rise in body temperature that is followed by a compensatory drop a few hours later. This decrease in body temperature, which persists for two to four hours after exercise, makes it easier to fall asleep and stay asleep.[45] The beneficial effect of exercise on sleep is greatest when exercise occurs within three to six hours of bedtime. Exercising closer than three hours to bedtime, however, may elevate body temperature too greatly, thus making it more difficult to fall asleep. So the best time to exercise if you want to sleep better is in the late afternoon or early evening.

Exercise also improves sleep because it is a physical stressor to the body. The brain compensates for physical stress by increasing deep sleep. Exercise may also improve sleep simply because people often exercise outdoors, which increases exposure to sunlight.

Stanford University School of Medicine researchers studied the effects of exercise on the sleep patterns of adults aged fifty-five to seventy-five who were sedentary and troubled by insomnia. These adults were asked to exercise for twenty to thirty minutes every other day in the afternoon by walking, engaging in low-impact aerobics, and riding a stationary bicycle. The result? The time required to fall asleep was reduced by one half; sleep time increased by almost one hour.[46]

Physical Activity Versus Intense Exercise

With all the benefits associated with exercise, why is it that more than 60 percent of Americans lead a sedentary life, getting little or no physical activity? Although people offer many reasons—the weather is bad, there is no good place to work out, someone has to watch the kids—perhaps the likeliest explanation is the misperception that exercise means sweat-soaked, exhausting, torturous physical exertion. This concept is partially the result of an overemphasis on the cardiovascular value of engaging in twenty to thirty minutes of high-intensity exercise three to five times per week, which has discouraged many. New scientific evidence indicates, however, that even moderate physical activity provides substantial health benefits. In 1996 the surgeon general recommended revised guidelines involving a gentler, easier exercise program of daily physical activity.

This program encourages adults to become and remain active by incorporating more movement into their daily routine. The guidelines recommend at least thirty minutes of moderate-intensity physical activity on most, if not all, days. This activity can come in the form of many of our normal daily routines: washing the car, taking the stairs instead of the elevator, or riding a bicycle instead of driving. It can be carried out in short segments broken up into several shorter sessions—say, of ten minutes each—that, during the day, add up to thirty minutes. That's enough physical activity to expend about two hundred calories daily. The total amount of activity is more important than whether it involves intense physical exercise or only moderately intense physical exertion. More recent evidence suggests that, for sedentary people, even one hour per week of light walking, where the time spent walking is more important than the walking pace, can reduce the risk of heart disease.[47]

Simply taking care of the house and the lawn qualifies as the kind of moderate physical activity that's good for you, as does playing actively with children, walking three to four miles per hour, or playing golf. A brisk walk, dancing, or fast bicycling will produce even greater health benefits.

Experts agree that most adults do not need to see their physician before beginning a program of moderate physical activity. However, it is a good idea to build up your endurance by starting with low-intensity activities for short durations a few times a week, then gradually increasing the duration and frequency to thirty minutes of total activity each day. If you plan to start a more vigorous physical exercise program or you have a chronic health problem, you should first consult your physician to plan a safe, effective program. Vigorous exercise should also be preceded and followed by a few minutes of stretching to work out muscle tightness and reduce the risk of muscle injury.

The fact is, we simply don't function as well without exercise. Movement, one of the oldest prescriptions for improving your health, is also the simplest and one of the most effective. Make it a regular part of your life.

Chapter Twelve

Solitude and Wilderness:
Medicine for the Soul

Humans thrive on variety. We like to try new foods, see new places, meet new people. Too little stimulation and we become bored; too little change and life becomes routine. But the hyperstimulation of the modern world—whether in the form of the media, advertising, faxes, beepers, cell phones, airplanes, traffic jams, or crowded commuter trains—threatens our health and well-being by depriving us of the calm and tranquillity that is our evolutionary birthright.

Our entire development as a species took place in an environment of relative calm, with the birds and the wind producing most of the ambient sound. But now, between the TM's endless chatter and the relentless assault upon our senses by the TM's technological handiwork, many of us not only have become habituated to overstimulation, but have become uncomfortable with quiet. Silence and solitude provide a context in which to experience the essence of our existence, but so many of us have become alienated from ourselves, numbed by noise and distracted by busywork, that the prospect of such an encounter with ourselves seems almost threatening. We've become deathly afraid of what we might find if we slowed down long enough to experience our own lives as anything other than a blur. So we accelerate the pace and keep the volume pounding.

Perhaps the biggest intruder on tranquillity and a healthy inner life is television. The medium has become an addiction, fueled by the forces of commerce, which see us only as eyeballs to be delivered to advertisers. Television invades our restaurants, stores, waiting rooms, airports, and health clubs, as well as every room in the home. It replaces real life with an endless stream of images and sensations, the vast majority of which are disposable and instantly forgettable. Instead of experiencing the richness and fullness of real life, we settle for a trivial counterfeit that steals our time and gives us nothing in return. As the title of Neil Postman's classic book suggests, when we watch television we are doing little more than *Amusing Ourselves to Death*.

In similar fashion, cell phones and pagers have the capacity to interrupt the flow and integrity of any real experience by rudely yanking us into someone else's concerns at any moment. These devices keep us constantly "in touch," yet on another level, they ensure that we stay *out of touch*—with ourselves and with the deeper, healthier world we might get to know within.

Accustomed to constant stimulation from media and from superficial encounters with others, we have lost the ability to be alone. In the process, we have lost the ability to be ourselves. As is the case with appropriate social connectedness, occasional solitude helps us cope with stress. Solitude can also be a great source of creativity, self-discovery, and spirituality. It is what we need at times to make sense of our lives, to integrate all the pieces, to make sure that we are the AM's whole person instead of the TM's fractured collection of social roles and obligations.

Many traditional cultures that are based on strong social networks also allow for formal opportunities for solitude, especially when young people come of age. These rites of passage include activities like walkabouts and vision quests, which have been carried over into the "solo" experiences offered by organizations such as Outward Bound. In traditional cultures ritual isolation is used to promote the appearance of spiritual dreams and visions, which are considered a critical component of preparing oneself to take full part in society. Young Native American men, for example, would go into the wilderness without food and water, build a sweat lodge, and pray for contact with a guiding spirit. From this transcendent experience, they might have visions in which they would discover their totemic animal, magic symbols, adult name, power, and direction in life.

Solitude not only enables us to assess, regulate, and change our lives, it also restores energy by allowing us to rest. It gives us a needed break from others that can also serve to strengthen relationships. Because solitude encourages us to get in touch with our deepest thoughts and feelings, it promotes healthier mental functioning. Solitude also aids people in coming to terms with loss, sorting out ideas, and changing attitudes.

Dr. Anthony Storr, author of *Solitude: A Return to the Self*, sees the need for social contact and the need for solitude as two opposing drives that operate throughout our lives. He considers the desire to be alone both a natural, evolutionary predisposition and evidence of emotional stability.[1] Storr believes solitude is biologically adaptive because it enhances imagination, and therefore is necessary if the brain is to function at its best and the individual is to fulfill his highest potential. Solitude likewise facilitates learning and innovation.

■ Solitude and the Mind/Body Connection

It's not surprising that solitude can have so many therapeutic effects on the mind and body, because in many respects it offers benefits similar to those of the relaxation response. It involves a reduction in sensory input that lowers arousal, and quiets the neocortex.[2] It produces states of mental quiescence that restore energy and promote positive mind/body interactions, but without the need to consciously use a repetitive mental stimulus to block sensory input. It allows us to open channels of communication with the AM and suppress the TM's normal rational thinking processes, and gives us access to stressful emotional material that has been registered unconsciously in the AM. Because of the reduced sensory input into the orientation association cortex and the prefrontal cortex, solitude can provoke alterations in consciousness and sense of self and time. And like the RR, solitude may enhance creativity and problem-solving because it enhances primary-process thinking, an almost childlike state of mental playfulness and profound absorption that allows ideas to be organized into new combinations and structures.

As an undergraduate, I investigated the effects of solitude on the mind and body by building a "sensory isolation" tank, in which by floating on saltwater in a quiet, dark, temperature-controlled environment, subjects could attain a deep state of relaxation. I found that the tank produced reliable reductions in blood pressure and muscle tension—the same effects produced by the other relaxation techniques.[3] As a placekicker on my college football team, I actually used the isolation tank myself to visualize and mentally rehearse my technique the nights before games. Two of my teammates, both running backs, also spent time in the tank to mentally rehearse their skills. The exercise seemed to work; I became a record-setting placekicker, and my two team-mates became all-Americans and received tryouts with professional football teams. It is likely that sensory-restricted environments not only enhance re-laxation skills but also enhance the efficacy of visual imagery techniques.

As a vehicle that can put the individual in touch with the possibility of mystical experiences, solitude affords an opportunity for personal explo-ration and growth and a sense of mastery. Writers as different as Walt Whit-man, Arthur Koestler, and C. S. Lewis all have left accounts of childhood feelings of mystical union with nature that occurred in solitude. Throughout human history, people have sought solitude in an effort to experience the presence of God. Mystics have often retreated to caves or wild places, and in the early Christian era some even sat for months atop tall poles. According to the Bible, Jesus spent forty days in the wilderness fasting and undergoing temptation by the devil before returning to proclaim his message of repen-tance and salvation. Mohammed and the Buddha also achieved their insights while isolated in the wilderness. Indian yogis and Tibetan monks still main-tain their mountain retreats and monasteries to develop enhanced control over the inner world of their Ancestral Mind through silence and solitude. Their experience is not of deprivation, but of unity, the feeling of the personal self disappearing, and the sense that the horizon of awareness has been greatly expanded. Esoteric traditions such as Zen use solitude as a means of enhancing their experience of the special instance of the Ancestral Mind, which they call *satori*, or enlightenment.

In the words of Thomas Merton, "It is in silence, and not in commotion, in solitude and not in crowds, that God best likes to reveal himself most inti-mately to men."[4] Again, prayer isn't so much a matter of *talking to* God, but of *listening for* God, which requires a willingness to embrace both silence and solitude.

■ Reexperiencing Solitude in Everyday Life

Seeking out solitude on a regular basis is an essential prescription for our overstimulated world. If you are practicing the relaxation response—settling in a quiet place, consciously reducing the sensory input into your brain—you have already begun to experience some of solitude's benefits.

Finding a little therapeutic solitude may be as simple as rising earlier each morning when the streets are still quiet, your family is asleep, and the sensory stimulation of the day is at its lowest ebb. Or perhaps you can steal away at lunchtime and spend a few minutes in a park or on a nature trail. Obviously, taking a weekend to hike in the mountains or go fishing or bird watching or canoeing will provide a deeper level of solitude that can give you even richer access to your Ancestral Mind. Whatever you can manage, it is more than worth the effort. Solitude and green space aren't a luxury, but essential for maintaining a healthy mind and body.

■ The Lost World of the Wilderness

In chapter five we described the ancestral world in which humanity evolved, a place where we were fully integrated into the natural environment. The deepest parts of ourselves are still profoundly connected to the wild—however much the TM's sense of detachment may interfere with the impulse— and we need to restore and maintain that connection to ensure our health.

Again, this claim is not just the expression of a romantic or sentimental or aesthethic preference. Science has demonstrated that simply viewing a natural scene can lower blood pressure. In a study conducted by psychologist Aaron Katcher and his colleagues, subjects' systolic and diastolic pressures were measured while they sat quietly, read aloud, and gazed at tropical fish in an aquarium. Katcher found blood pressure levels were highest when reading aloud, and slightly lower when resting in a chair. But the lowest readings came when subjects gazed at the tropical fish. Interestingly, when asked to read aloud after watching fish, subjects experienced a rise in blood pressure less than half as large as the "reading aloud" level at the start of the experiment.[5] Katcher's conclusion: that the calm induced by watching fish reduced subjects' response to the stress of public performance. Viewing nature draws attention outward, interrupts the flow of thoughts, and produces states of reverie like those we've seen in the relaxation response or stage 1 sleep, all of which, we know, lessen our preoccupation with the self.

In another study published in the journal *Science*, psychologist Roger Ulrich found that patients who underwent gallbladder surgery and had hospital rooms with windows that looked out onto trees spent less time in the hospital after surgery, were less upset, and took less pain medication than patients whose windows looked out onto a blank wall. Ulrich also found that people in stressful situations who viewed slides of nature as compared to scenes of buildings showed lower stress responses and faster physiological recovery from stress.[6]

Rooted as it is in the biological makeup of the Ancestral Mind, nature triggers genetic memories of the world in which we evolved. In his book *Biophilia*, National Medal of Science and two-time Pulitzer Prize winner E. O. Wilson makes the case that we prefer natural landscapes over manmade ones as a direct result of our evolution. When people view slides of nature scenes, they report higher levels of positive emotions than when viewing urban scenes. Not only do we find natural scenes preferable, but we are intuitively attracted to certain landscape configurations because they were desirable from an evolutionary perspective.[7] Universally, humans prefer scenes with smooth ground cover, scattered trees and lakes, and openness and depth, probably because these are the characteristics of the savannas of Africa, where our ancestors evolved. Trees and water, in particular, enhance positive emotions, presumably because ancestral environments that lacked these features were less likely to support human life.

Biologist George Orians and his colleagues support this view that our present-day sense of natural beauty is the reflection of forces that drove our ancestors into suitable habitats.[8] For example, we innately find savannas beautiful because, in ancestral times, this habitat contained key features necessary for survival: semiopen spaces (neither completely exposed, which leaves one vulnerable to predators, nor overgrown, which impedes vision or movement); views to the horizon; large trees, both for protection from the sun and for shelter, to be climbed to avoid predators; adequate food and water; changes in elevation that allow orientation in space; and multiple paths leading out.

Rivers and mountains are calming because they act as landmarks; a vista without such topography can seem unstructured and therefore unsettling. Yet another key to our sense of natural beauty is mystery. Paths that bend around hills, streams that meander, undulating land, and mountain ranges capture our interest by hinting that the land may hold appealing features that should be explored. Sunsets, thunder, or gathering clouds hold our attention because they foretell change: darkness or a storm. We are attracted to flowers not only because they are beautiful but also because they flourish in areas that mark the site of foods, such as fruits and nuts.

Natural habitats that provide a view and a sense of refuge and mystery activate the neural pathways of positive emotions in the AM because they were associated with what evolutionary psychologists term "fitness"; that is, in ancestral environments, choices of such settings contributed to greater health, well-being, and survival.[9] Today we may experience positive emotions in response to a variety of natural stimuli in very different environments from the ones in which we originally evolved, but we are still activating ancestral emotional responses that served important functions in both day-to-day survival and the long-term well-being of early humans.

Nature also allows us to retreat and disappear from the troubles of life by radically altering our perspective. It arouses a deep sense of familiarity with our ancestral past. The vast diversity of life on this planet was here millions of years before our own brief existence as an individual; presumably, it will be here, rolling on as before, long after each of us is gone. The nighttime sky, for example, has been a source of beauty and inspiration, even mystical awe, since the beginning of human history. It's no wonder that it has so often been used as the basis of religious rituals, myths, and astrological visions. Few of us have looked up at the billions of stars and not been left breathless by their spellbinding power, yet how often are we really able to fully enjoy this experience? With the profusion of lights in our manmade world we have driven away the vast serenity, austerity, and mystery of the stars. Light pollution leaves us with an empty, night-lit desert, a darkness that is little more than artificial dusk, a dusk that is turned off by dawn.

"Mountains are the beginning and the end of all natural scenery," wrote John Ruskin, England's first professor of art. Mountains have obvious dominion over the land and nature. They catch the first and last rays of sunlight, and allow us to discover horizons we have never seen. In the mountains, the mind and senses become clearer. For believers, they are the closest points on the planet to the home of the gods, connecting the spiritual world with our own. Even the most rugged alpinist returns from the mountains filled with an inexplicable sense of inner peace.

For most of human history we were inseparable from nature; we *were* wild nature. It was only about ten thousand years ago that we began to tame flora and fauna. Then came towns and cities, followed by industrialization, which has, in the last few centuries, separated most of us from the experience of the wild. We have turned nature into something utilitarian, and we have lost the deeper connection to it as our spiritual home. We have become disconnected from our natural rhythms, our link to the past, and our most powerful link to the Ancestral Mind.

Thoreau spoke of "the tonic of the wilderness," by which he meant its power to take us outside of ourselves and reconnect to the beginner's mind, the child's mind, in which the world still appears fresh, unbounded, magical, and alive. American Indians honor nature as a divine source of life and believe it has a soul. Its infinite dignity allows us to reclaim the permanent to balance the modern. Nature is a visceral reminder of a sense of continuing creation and the drive for life. In the wilderness, we can lose our everyday TM selves, and thereby find our deeper AM selves.

How do we show our appreciation for the gift? By clear-cutting, bulldozing, and blacktopping nature. Today one in every four mammals and one in every eight birds is currently at risk for extinction. Conservationists estimate that the current extinction rate is one thousand to ten thousand times higher than it should be under natural conditions. The primary reasons: the TM's urbanization, deforestation, agriculture, and commercial fishing.[10] We would be horrified if we had to face the destruction of our cultural heritage, the great works of art of many centuries, yet we are allowing the very same thing to happen to something more precious, and a vital part of what makes us human.

Each year I return to my favorite western river to fly-fish, and it is my time there, more than anything else, that enables me to connect deeply to the more grounded part of myself. I have stood in the same spot on this river year after year and have come to know the area's mountains, trees, rocks, birds, and deer. They have become friends that link me to timeless rhythms like those my ancestors were a part of long ago.

On this river I have seen the sun and the sky in ways that I have never seen them before. It is here, after several days of solitude, that my sense of time and self is transformed, my mental conversation silenced. I become immersed in the moment. The river feels so alive that I find myself engaged in a wordless dialogue with it. I am no longer concerned with what things *mean*; I am concerned only with what *is*. I begin to see new relationships and patterns that redefine "knowing." I realize that my scientific mind cannot explain the beauty and mystery of what I see. I let the river into my mind, hear its voice, feel its life, so that I am part of the river, and it is part of me. The more I learn about the river, the more I learn about myself.

The TM tells us that rivers and mountains and the nighttime sky can be explained in scientific terms, but the AM tells us something else—that they are full of magic and mystery, and that they are our home.

■ Reexperiencing Nature

Although we may ideally wish to celebrate nature in remote and glorious places, it can also be experienced on a smaller scale nearer at hand. We've already described the therapeutic effects of gazing into an aquarium, or even a landscape painting. In the same way, we can benefit from watching a spider weaving its web, a bird building its nest, or just by listening to the sound of rain or the wind soughing through a pine grove. We can garden, take a walk in the woods or a park, get involved in photography, or listen to audio nature recordings when we feel stressed.

Often the beauty of nature is immediately present before us, but we are simply too preoccupied to appreciate it. We overlook the beauty of a flower or a sunset because of the distraction of the internal monologue. We drive to work and don't notice a magnificent sunrise or the splendor of the leaves changing color on a sunny fall day.

Pets, too, especially mammalian pets, reconnect us to the AM, as well as providing social support and affection. It is likely that the love and companionship that we feel for a pet activate health-enhancing pathways in the AM.

■ Medicine for the Mind and Body

Like the relaxation response and solitude, nature can have profound neurophysiological effects on the AM:

• Because stunning fall foliage or a magnificent autumn moon are intrinsically pleasurable and calming to the brain, nature can activate the neural circuitry of positive emotions.

• By commanding our attention, absorbing us, and providing silence and stillness, nature alters attention. The more constant, natural sensory stimulation of nature acts as a potent mental focus that reduces arousal in the reticular formation, shuts down information flow from the thalamus to the neocortex, and quiets the orientation association cortex, the prefrontal cortex, and working memory.

• Nature elicits an ancestral state of mind: one of wonder and amazement, awe and transcendence. By profoundly entrancing us and taking us outside of our everyday existence, nature helps us connect to feelings of something greater than ourselves.

All our lives we've heard expressions such as "Don't forget to stop and smell the roses," and modern melodrama has made a cliché of sentiments such as "I just need some time for myself." But don't let familiarity obscure this essential fact: nature and occasional solitude are powerful therapeutic agents of proven benefit to your health. Take them seriously, and weave them into your effort to reconnect with the Ancestral Mind.

Chapter Thirteen

Through the Eyes of a Child

This book began with a series of questions:

Why are we so unhappy? Why is there so little satisfaction in our lives? Why is there so much frustration and stress?

The cause we have identified to account for our condition is the overshadowing of the Ancestral Mind by its self-absorbed cousin, the Thinking Mind.

In the preceding pages we have examined the scientific research that supports the existence of and significance of this more ancient but more grounded Ancestral Mind. We have discovered that it is the control center that has watched over us throughout the course of human evolution—preverbal, directly connected to the real world through the senses and the emotions, engaged only in the here and now. Most important, we have also seen that it is endowed with an intuitive knowledge, gained through millions of years of natural selection, of what we truly need for health and well-being.

By explaining the underlying science we have tried to make it clear that this distinct, separate, and generally neglected mind is no mere metaphor but something quite real, and clearly associated with specific anatomical structures in the brain. When we reconnect with this ancient part of ourselves, we alter the neural activity in certain structures in the brain, and that, in turn, can lead to a greater sense of peace, contentment, and a more intense connection to all the things we value most in life.

Neuroscientist Joseph LeDoux suggests that resolving the struggle between Thinking Mind and Ancestral Mind is actually a matter of integrating the two. When we strike a greater balance between thought and emotion, we may actually promote increased neural connectivity between the neocortex and the mammalian brain, enabling cognition and emotion to reinforce, rather than work against, each other. In that way, the Ancestral Mind not only promotes physical and emotional well-being, but enhances mental clarity, energy, and concentration as well.

Renewing this healthful connection involves learning how to overcome the preoccupations of the Thinking Mind, a process that involves moving beyond the boundaries of self and object, the obsession with past and future, fears and regrets—all the mental habits that lead to our endless cycles of work and worry.

The healing attitudes and the mind/body techniques we have explored serve to further that more harmonious integration of the TM and AM. The use of these techniques to strike a greater balance in the neural pathways between the AM and TM may actually represent the next stage in the development of humanity, one that enables us to become consciously involved, moment by moment, in promoting our own health and well-being.

To be effective, though, these habits and techniques can't be something you simply read about: they have to become a part of you. In the grander scheme of things, for some people, this could mean making significant changes in career or location or lifestyle. For others, it might involve seeking the help of a stress reduction clinic, making a conscious decision to shift the balance between work and a meaningful avocation, or simply signing up for a yoga class. At the very least, it should include incorporating a new outlook and new behaviors into your daily routine.

No matter how you translate the desire to reconnect with the AM back into your own life, the basic message can be summarized in a few guiding principles that can serve as compass points to keep you on track, and as touchpoints to inform your choices. These principles, in their simplest form, are the need to:

- Trust emotion and the unconscious as well as reason and conscious awareness
- Value and take time for the simple pleasures that produce positive emotions

- Tame negative automatic thoughts through cognitive restructuring and reframing
- Allow yourself to be in the moment, receptive to states of flow, as well as to peak experiences (including, perhaps, mystical experiences)
- Value and take the time for social connections
- Pursue altruism, cooperation, and optimism, not only because they are noble qualities, but because they are healthful and adaptive qualities
- Use guided imagery, affirmations, and an attitude of gratitude to develop these stress-reducing qualities
- Reduce anger and hostility through the eleven steps we discussed in chapter 7
- Replace negative emotions with the "natural high" of laughter
- Pursue positive illusions, even positive denial
- Promote stress hardiness through control, commitment, and challenge
- Take time for reverie, for naps, and for a good night's sleep
- Practice the relaxation response
- Insist on getting your minimum daily requirement of music, light, and exercise
- Indulge your need for wilderness and solitude
- Develop faith, and allow yourself a sense of mystery and wonder.

But there's an even simpler way to capture the essential message of the Ancestral Mind. Throughout this book we have referred often to the child's sense of wonder, the child's enviable appreciation of play and lack of self-consciousness, the child's delightful ability to be fully absorbed in the moment. The fact is, the child's state of being is the best exemplar we have of the Ancestral Mind.

The spontaneity and self-forgetfulness of children enable them to experience a more immediate sense of the enchantment in everyday moments. Children play for the sheer joy of it and have a natural curiosity about everything. They can change their perspectives quickly and dramatically, and are remarkable observers; they put us to shame with their ability to perceive particular details that we are too lost in thought to notice. They have conversations with their toys, enjoy their own good company, and focus their attention with breathless concentration. A child at play is the epitome of someone living in the present.

According to the existential psychologist Dr. Erich Fromm, a very young child's experience of the world is more immediate and direct, simple and spontaneous because he perceives reality without "intellection," which, in our terms, means that he lacks the distorting influence of the Thinking Mind's internal monologue. Like preverbal humans tens of thousands of years ago, a child does not *think* about an experience; the child is fully *in* the experience. It is only with the coming of words and the pressures of social conventions that the child learns to detach, objectify, and analyze.

The eminent child psychologist Dr. Jean Piaget also described children as not being conscious of themselves as a thinking self. They do not regard themselves as observers, set apart from the external world. Their boundaries are less rigid. They feel themselves to be identical with the images they perceive—they *are* the world, focused not on themselves but on the things around them.

According to Piaget, the child's worldview is also characterized by animism (the belief that inanimate objects have consciousness) and a sense of magical participation in commonplace events. Since there is no distinction between self and the external world, everything participates in the nature of, and can influence, everything else. Like ancestral humans, children endow material things with feelings and consciousness; they believe thought to be inseparable from an object, names inseparable from the things named. Thoughts, images, and words are situated *in* things, as if the mind and the thing are one.

In *The Re-Enchantment of Everyday Life*, Thomas Moore describes the child's sophistication prior to becoming an adult, which, in Moore's words, involves an appreciation "of the interiority of nature and the hidden power of persons and places." In their early years, children believe their actions and thoughts can magically alter external events and modify reality; they see human causes for natural events and see human attributes in inanimate objects. When children close their eyes, they cause the world to disappear; when they open their eyes, they cause the world to reappear.

In our discussion of Dr. Shelley Taylor's research, we learned that positive illusions are present in a strong, almost magical degree in children. Taylor believes that, far from being primitive, pathological, and irrational, these positive illusions are likely evolutionary accommodations that are necessary for emotional health.

As we grow up, we fall out of and forget this enchanted world of childhood. We lose the sense of wonder and mystery, the dreams and imagination; we stop seeing the animals in the clouds, our positive illusions disappear, our natural joy and wild-eyed optimism fade as we are educated in the ways of the TM. Science and rational thought become our arbiters of reality, and we lose our connection with other ways of experiencing the world. We become more analytical and serious and learn not to believe without evidence. That rational skepticism has its place, of course; the tragedy of modern life is that objectivity and reason do not develop in us to complement our sense of mystery and wonder, but to replace it. The better course, and the key to mental health, may be in resisting an all-encompassing rationalism, and as Shelley Taylor advises, balancing reason with a sense of the magical world of the child that promotes a positively biased, "unrealistic" view.

Many of us remain nostalgic for those early years, because we sense at a deep level the joy and serenity we have lost. There exists a place deep within us that still holds the old idols, the old beliefs, and the old magic. Every so often, when we are reminded of how we looked at the world when we were young, we get a glimpse into the Ancestral Mind.

There is a common theme that unites the research of Piaget, Fromm, and Taylor: If we can minimize our sense of self and a preoccupation with objectivity and rationalism, we can reawaken the enchantment of childhood, which means reawakening our connection to ancestral states of mind. Although we have outgrown our childhood, we can nonetheless become more childlike.

The goal of learning to perceive the world again as children is at least as old as Christianity. In the New Testament we find: "Verily, I say unto you, Whosoever shall not receive the kingdom of God as a little child, he shall not enter herein" (Mark 10:15). Jesus also said, "Except ye be converted, and become as little children, ye shall not enter into the kingdom of heaven" (Matthew 18:3). Two thousand years later, speaking from a secular, psychological perspective, Erich Fromm echoed the same sentiments, observing, "We have to become like children again, to experience the unalienated, creative grasp of the world; but in becoming children again we are at the same time not children, but fully developed adults."

As adults, we must find time to appreciate the small pleasures in life, and we must give ourselves permission to do "nothing," to play. Play renews us, enhances feelings of youth and vitality. But we may need some help in remembering how. The first step could be in becoming careful observers of children, who can teach us the skill of being present in the moment. Children can give us an excuse to have fun and look outside of ourselves. Merely watching children can trigger positive emotional memories of our own childhood and the thrill of play, games, toys, laughter, and songs long since forgotten.

To follow Fromm's advice, to reacquire the mind of a child after developing the TM's cognitive powers, is to regain childhood on a higher level, and a new level of well-being. Eastern philosophies characterize the TM-based consciousness of adulthood as being "half-sleep." Finding the proper balance of Ancestral Mind and Thinking Mind would mean living a life that is fully awake to all its possibilities.

Appendix A
Physiology and the Brain

Neurons and Neurotransmitters

The three brains—reptilian, mammalian, and neocortex—consist of approximately 100 billion cells called *neurons*. These basic units of the brain carry tiny bursts of electrochemical energy that travel from neuron to neuron across tiny junctions called synapses. Each neuron is connected via a synapse to thousands of other neurons, forming chains of neural pathways that resemble strings of Christmas tree lights. A single thought involves an electrochemical signal that travels along millions of neurons and perhaps trillions of synapses.

The chemical messengers that enable neurons to communicate with one another are called *neurotransmitters* (NTs). The NTs that are important in emotions include serotonin, epinephrine and norepinephrine, dopamine, GABA, and the endorphins. Some of these activate brain structures (that is, they are excitatory), while others are inhibitory and quiet brain structures. The two NTs that are most central to mood are serotonin and norepinephrine. Serotonin quiets aggression, relieves anxiety, and enhances mood by calming the amygdala. Antidepressants such as Prozac exert their mood-enhancing effects by increasing supplies of serotonin in the brain. Norepinephrine is involved in attention and alertness and can help us to feel less fatigued; it is also involved in the stress response. GABA, an inhibitory NT, is also important to mood because it reduces anxiety by deactivating the amygdala. Many antianxiety medications like Valium exert their calming effects by increasing GABA concentrations in the brain. Dopamine is involved in feelings of pleasure.

Endorphins, which are the brain's own endogenous opiates, are also involved in pleasure and may play a role in feelings of love and attachment. Some hormones, such as oxytocin and prolactin, have also been linked to emotions. Oxytocin plays a role in sexual activity and feelings of attachment and love. Prolactin is secreted by mothers when they breast-feed and may be responsible for the calm, serene state that they feel during this nurturing act.

The Reptilian Brain

The reptilian brain, or brain stem, performs two basic functions. Its upper part receives incoming sensory messages from the senses and sends them upward to the thalamus; its lower section sends motor commands to the muscles, organs, and glands. Through the reticular formation, the reptilian brain also plays a crucial role in arousal. We could not have consciousness without the reptilian brain.

The Mammalian Brain

The mammalian brain consists of numerous brain structures, including, among others, the amygdala, hippocampus, cingulate gyrus, thalamus and hypothalamus, and parts of the prefrontal cortex. Technically, the thalamus and hypothalamus can be considered part of the reptilian brain since they are older than the mammalian brain, but because they are so highly developed in humans and play such a major role in emotion, they can be regarded as belonging to the mammalian brain. The mammalian brain grew out of the olfactory bulb of the reptilian brain and was originally called the "smell brain." It is made up of two layers of evolutionarily older limbic cortex that lie just beneath the newer neocortex, which itself grew out of the limbic cortex. Limbic cortical areas such as the cingulate cortex receive direct input from the neocortical association areas.

The mammalian brain evaluates and integrates all of the information that travels between the reptilian brain and the neocortex for emotional significance; it also directs emotional behavior by sending signals to the reptilian brain to initiate basic emotional responses like fight or flight and to the neocortex for the emergence of feelings and facial expressions. Because the mammalian brain is the seat of emotional consciousness, it is the engine that drives motivation through emotions.

The thalamus constructs and elaborates sensory information entering the brain and then relays this information along two roads: a low road to the amygdala and a high road to the neocortex. These two roads are called the thalamoamygdala and thalamocortical pathways, respectively. Not all sensory stimuli travel the thalamocortical pathway and register in the neocortex and conscious awareness because the reticular formation may not generate enough arousal to allow this to occur. The hypothalamus receives direct sensory and emotional inputs from the thalamus, controls bodily responses during emotion, and communicates with the neocortex. Because of its significance in emotion and motivation, the hypothalamus connects to all parts of the brain.

The amygdala receives sensory information from the thalamus along the low road, integrates the information somewhat crudely, processes the information for emotional importance, then transmits signals to the brain stem and neocortex. Because the amygdala sends out connections to virtually all parts of the brain, it can influence thought, perception, memory, movement, and stress responses; it also affects arousal and concentration by directing the reptilian brain to produce more epinephrine and dopamine. The amygdala also receives messages from the cortex, which allows the amygdala to monitor emotional stimuli from both lower and higher brain regions.

When the amygdala deems emotional stimuli meaningful or dangerous, it trips the hypothalamus to initiate the stress response automatically. The amygdala also instructs the reticular formation to generate sufficient arousal to ensure that important emotional stimuli reach the cortex and enter conscious awareness in working memory in the prefrontal cortex (PFC). Sensory stimuli are then elaborated and modified by the PFC, which modulates the amygdala's response.

The hippocampus is involved primarily in memory. However, because its purview is factual but not emotional memory, it does not play as salient a role in emotions as the amygdala.

The mammalian brain and neocortex represent somewhat distinct neural systems. They speak different languages—one by feelings, one verbally—which is why they are sometimes referred to as the "emotional brain" and the "word brain." Indeed, Dr. Paul Maclean's theory of the triune brain contends that each brain has its own sense of time and space. For these reasons, the two don't always communicate well. The neocortex has a hard time comprehending feelings, intuition, and another way of knowing—the reasoning of the heart. There are reciprocal neural connections between the two brains, but the connections from the mammalian brain to the neocortex are stronger than vice-versa, which gives the mammalian brain the ability to exert greater influence over the neocortex in emotional matters.

The Neocortex
The neocortex synthesizes what the senses perceive and puts the information together to allow us to comprehend the world, have ideas, use symbols, and imagine. Specific areas of the neocortex are involved in specific functions. For

example, visual processes are localized in the occipital cortex, language comprehension in the temporal cortex, touch and movement in the parietal cortex, and attention and awareness in the frontal cortex. However, because these areas are dependent upon other brain structures to perform these functions, localizing brain functions to specific brain regions should not be taken too literally.

When sensory information enters the brain, it travels to primary cortical areas devoted to that particular function and is then integrated and synthesized into complex thoughts and perceptions—the building blocks of consciousness—in association cortices (the secondary cortex). Ultimately, these association cortices tap memory and emotional information to create complex perceptions of the world and conscious thought.

It is important to note that many areas of the cortex involve the complex motor or sensory processing and execution of tasks such as language that are not conscious. For example, much of the sensory information that arrives at the primary cortical areas such as the visual cortex are rough visual impressions like lines, shapes, and colors that aren't consciously perceived but are then integrated into complex conscious perceptions. Thus, equating "conscious" with the neocortex is not accurate.

The neocortex consists of two hemispheres. Whereas the left hemisphere specializes in language, mathematics, and analytical, sequential tasks, the right has its own form of nonverbal awareness and specializes in spatial tasks and the perception of distance and music. However, the two hemispheres draw off each other. Most higher-order mental functions involve the participation of both hemispheres, although not equally, so that consciousness is the result of the integrated workings of both.

The parietal cortex and its specialized area, the orientation association cortex, mediate behavioral interaction with the world around us. They are essential for a complete self-image and for determining where objects are located in space. One of the characteristics of the baseline state of the human brain is activation of the parietal cortex due to attending to and interpreting external stimuli in relation to the self. The left orientation cortex creates a subjective sense of self, while the right generates a spatial context in which the self orients itself.

The PFC performs many complex functions and consists of several regions. It plays an important role in synthesizing information from other cortical areas, coordinating the functions of these areas, and organizing information from other cortical areas into perceptions and ideas. The PFC also gathers, organizes, and integrates information from the reptilian and mammalian brains. Consequently, the PFC has more connections to other parts of the brain than any other cortical area.

Due to its extensive reciprocal connections with the amygdala, the PFC plays a role in planning and carrying out emotional actions; it organizes perceptions in the pursuit of goals. The PFC modifies impulsive, rapid emotional responses generated by the mammalian brain by modulating the responses and arriving at a more appropriate response. The PFC does this by assessing the emotional stimulus, considering and weighing options, and then planning and initiating an emotional response by sending signals back to the reptilian and mammalian brains. In this fashion, the PFC allows for more flexibility in our emotional behavior. Several areas of the PFC, including the medial, anterior cingulate, and orbitofrontal areas, are involved in emotional behavior through their connections with the amygdala. Some of our most sophisticated cognitive functions, such as predicting outcomes and prioritizing actions, also occur in other regions of the PFC, especially the lateral and dorsolateral regions.

Consciousness and its attributes, such as imaginative constructing of scenarios, hypothesizing, theorizing, and symbolizing, are dependent upon working memory. Working memory (WM) is akin to a buffer or type of temporary storage space in the PFC that allows us to hold information and manipulate it in an abstract fashion. WM is an active processing mechanism used in focused attention, thinking and feeling, and reasoning so that we can plan and make decisions.

WM is crucial in the ability to reflect on our own perceptions. It allows us to hold several different thoughts in mind at the same time so that we can go offline from the present in order to combine several thoughts together in an abstract fashion to form symbolic representations about the past, present, and future. WM consists of the things we are currently paying attention to or thinking about. In this sense, consciousness is awareness of what is in working memory. Feelings, for example, are the conscious awareness of emotions that have registered in WM.

WM is linked to long-term memory and short-term memory. Because much of what we recognize and think about is based on past memories, WM consists of an interplay between long-term memory and information currently in short-term memory. WM gives us the ability to monitor our own thoughts and behaviors in relation to the past and future. A central feature of WM is language, which enables us to symbolically categorize and label experiences, and to create an abstract symbolic representation of the world. Without this ability, conscious thought would be impossible.

WM allows us to project ourselves forward and backward in time, which is crucial to a sense of self. It is in WM that we become aware of our self in relation to time and place. The ability to recognize one's self in the mirror depends upon the ability to hold an image of one's self in WM, link the image to past images and memory, and consciously have the thought "that's me." Without WM, there would be no memory of a personal history, no sense of past or future and therefore of context and continuity for the self. To lose or alter sense of self, time, and the internal monologue means to alter mental processing and information content in WM.

Appendix B
The Effects of Stress on Health

Stress involves a complex set of biological and chemical responses that begin in the brain when demands are placed upon an organism. Under exposure to stress, the hypothalamus releases a hormone called CRH, which signals the pituitary gland to secrete a hormone called ACTH into the bloodstream. ACTH in turn encourages the adrenal glands to secrete hormones into the bloodstream, including cortisol, cortisone, and the catecholamines (epinephrine and norepinephrine). Catecholamines stimulate the autonomic nervous system (resulting in increased cardiovascular, respiratory, and skeletal muscle activity; greater alertness and energy; increased blood flow to the heart and lungs; and suppression of eating and sexual activity) and arouse various brain regions, including the brain stem, amygdala, and prefrontal cortex. Although these neuroendocrine changes are part of the nervous system's normal adaptive responses to stress and protect the body, the same changes can cause excessive wear and tear on the nervous system if they occur too frequently. Excessive stress responses can provoke disease over time, leading to cognitive, emotional, and physiological disturbances.

Various studies have shown that stress impacts immune functioning through links between the central nervous system and the immune system. Examples of these links include extensive networks of nerve endings that innervate the thymus gland of the immune system and rich supplies of nerves that innervate the spleen, bone marrow, and lymph nodes. Additionally, cells in the immune system appear to be equipped to respond to chemical signals from the nervous system. Specifically, the surfaces of lymphocytes contain receptors for a variety of hormones such as epinephrine that are mediated by the nervous system. These immune–nervous system connections allow stress to influence resistance or susceptibility to infectious or autoimmune diseases and possibly speed the metastasis of cancer. Stress also impacts immune functioning because both cortisol and the catecholamines can suppress the immune system; high levels of cortisol can even shrink the spleen and thymus gland. Whether or not the effects of stress on the immune system are great enough to be of clinical significance is not yet known, but the relationship between stress and immune functioning is convincing.

Cardiovascular Disease

When it is released into the bloodstream under stress, cholesterol can build up and cause heart disease. Blood pressure rises under stress (chronic high cortisone levels stimulate the kidneys to produce renin, a hormone that raises blood pressure), the coronary arteries constrict, and the blood thickens more readily, which can lead to hypertension and heart attacks. Stress can also cause sudden cardiac death through the effects of irregular heart rhythms caused by catecholamines. Stress compromises the pumping efficiency of the heart and can cause silent ischemia, a condition in which blood does not circulate adequately to the heart. Studies have shown that even low levels of stress commonly experienced in daily life are sufficient to trigger heart attacks in people with existing heart disease; emotional stress, particularly anger, fear, anxiety, bereavement, and depression, precedes both fatal and nonfatal heart attacks in about 15 percent of cases. Stress also contributes to heart disease by exacerbating plaque formation, which leads to atherosclerosis.

Hostility can damage the heart and may be as strong a predictor of mortality as smoking, hypertension, or high cholesterol. Hostility may play a role in the development of heart disease or may exacerbate heart disease once it begins; studies have shown that anger also increases the risk of heart attacks. When someone gets angry repeatedly over many years, the chronic increases in heart rate and blood pressure strain the heart. That's why hostile people are seven times more likely to die by age fifty than less hostile people.

Gastrointestinal Disorders

Stress alters the balance of the sympathetic and parasympathetic branches of the autonomic nervous system. There is a slowing of gastrointestinal motility under stress because blood is sent to the muscles, heart, and lungs for fighting and fleeing. These changes can lead to indigestion, abdominal cramping and pain, nausea, and irritable bowel syndrome. Stress can stop the contractions of the esophagus or can cause irregular contractions. Stress also appears capable of disturbing the peristaltic action of the entire gastrointestinal tract, which can lead to either constipation or diarrhea. Stress has been shown to cause ulceration of the gastrointestinal tract, which triggers ulcerative colitis.

Other Health Problems

Stress has consistently been linked to insomnia and disturbed sleep as well as headaches. Stress can exacerbate various types of pain, particularly chronic pain. Because excessive cortisol can lead to progressive nerve loss in the hippocampus (a brain region that is important in memory), stress may play a role in the development and progression of Alzheimer's disease. Excessive cortisol production causes the liver to overproduce glucose, which can increase blood sugar levels. This is why stress has been associated with accelerating the onset of Type I diabetes and affecting the course of Type II diabetes and has been shown to raise blood sugar levels in diabetic patients. Stress alone does not cause diabetes; stress interacts with other factors such as diet and genetics and is more likely to aggravate diabetes. An association between severe emotional stress and asthma has also been established, and stress has been implicated in infertility.

People who are chronically anxious, pessimistic, or hostile have double the risk of many diseases, including asthma, arthritis, heart disease, ulcers, and headaches. The doubling of health risks for these diseases makes the effects of stress on health roughly comparable to more traditional risk factors such as smoking or high cholesterol. People who experience the stress of social isolation also exhibit a doubling of morbidity and mortality rates, which rivals the effects of a sedentary lifestyle.

Stress can cause or make worse virtually all psychiatric disorders, including anxiety, depression, panic disorder, obsessive-compulsive disorder, and post-traumatic stress disorder. Stress can play a direct role in the onset of these diseases or exacerbate them over time.

When considering the far-ranging effects of stress on health, it is important to realize that stress may play a role in the development of many health problems but is probably not the sole cause. For example, people under stress are also more likely to drink alcohol, eat and sleep poorly, exercise less, and smoke cigarettes or use drugs, all of which can interact with stress to cause health problems. In the case of the immune system, stress may be more likely to enhance susceptibility to illness in those individuals who already have weaker immune systems, such as the elderly. In other cases, stress may not play a causative role in a health problem but instead may exacerbate it.

Suggested Readings on the Stress-Illness Connection

Ader, R. *Psychoneuroimmunology.* San Diego: Academic Press, 1990.

Cohen, S., et al. "Psychological Stress and Susceptibility to the Common Cold." *New England Journal of Medicine* 325 (1991): 606–12.

Felten, D., et al. "Noradrenergic Sympathetic Innervation of Lymphoid Tissue." *Journal of Immunology* 135 (1985): 755–65.

Friedman, H., and S. Boothby-Kewley. "The Disease-Prone Personality: A Meta-Analytic Review." *American Psychologist* 42 (1987): 539–55.

Goleman, D., and J. Gurin. *Mind/Body Medicine.* New York: Consumer Reports Books, 1993.

Hafen, B., et al. *Mind/Body Health.* Boston: Allyn and Bacon, 1996.

Maier, S. B., et al. "Psychoneuroimmunology." *American Psychologist* 49 (Dec. 1994): 1004–17.

McEwen, B. "Protective and Damaging Effects of Stress Mediators: Central Role of the Brain." In *Progress in Brain Research* 22 (2000): 25–33.

McEwen, B., and E. Stellar, "Stress and the Individual: Mechanisms Leading to Disease." *Archives of Internal Medicine* 153 (1993): 2093–2101.

Somerville, P. D., et al. "Psychological Distress as a Predictor of Mortality." *American Journal of Epidemiology* 130 (1989): 1013–23.

Verrier, R., and M. Mittleman. "The Impact of Emotions on the Heart." In *Progress in Brain Research* 22 (2000): 369–80.

Williams, R. *The Trusting Heart.* New York: Times Books, 1989.

Appendix C
Additional Relaxation Scripts

Autogenic Training

Autogenic training is a relaxation technique that was developed in Germany by Dr. Johannes Schultz. Autogenic means "self-induced"; autogenic training self-induces relaxation by utilizing suggestions of warmth and heaviness. The following script is a modified version of Dr. Schultz's original autogenic training:

Close your eyes and direct your attention to the toes and feet. Feel a wave of relaxation begin to move through the feet.

Now the relaxation moves through the calves and thighs, then to the stomach, chest, and back. Note tension disappearing from your body. Sense relaxation moving now to the hands, forearms, upper arms, and shoulders. Next the relaxation moves to the neck, jaw, cheeks, eyes, and forehead. Tension is disappearing from the body.

Focus your attention on your breathing. Note that the breathing is becoming more rhythmic and abdominal. As you inhale, the stomach expands; as you exhale, it contracts. Inhale, it expands; exhale, it contracts. Take a few moments to concentrate on abdominal breathing. You are breathing more rhythmically and abdominally.

If your mind wanders to distracting thoughts, gently let the thoughts go and return your attention to the breathing. It may help to repeat a word silently to yourself each time you exhale, such as the word "one" or "relax." This word will serve as a mental focus to direct the mind away from distracting, everyday thoughts. Take a few moments to concentrate on the breathing and the word.

Now that your breathing has become more rhythmic and abdominal, direct your attention to the arms. As you focus on the arms, repeat slowly to yourself "My arms are heavy." It may help to visualize an image of the arms becoming heavy. Slowly repeat "My arms are heavy" two more times. Direct your attention now to the legs and slowly repeat to yourself "My legs are heavy." Again, it may help to visualize the legs becoming heavy. Slowly repeat "My legs are heavy" two more times.

Concentrate your attention now back to your arms and repeat slowly to yourself "My arms are warm." Visualize an image of the arms becoming warm. Slowly repeat "My arms are warm" two more times. Focus next on the legs and repeat slowly to yourself "My legs are warm," visualizing the legs becoming warm. Slowly repeat "My legs are warm" two more times.

Your entire body is heavier, warmer, and more relaxed. Take a few moments to note the absence of tension throughout the body. You feel calm, relaxed, and peaceful.

Finally, at your own pace, take a slow, deep breath and slowly open your eyes.

Progressive Relaxation

Progressive relaxation is a technique that was developed by Dr. Edmund Jacobsen and described in his book *Progressive Relaxation*. It involves tensing and relaxing the muscles throughout the body to increase awareness of tension and relaxation and elicit the relaxation response. The following script is a modified version of Dr. Jacobsen's original progressive relaxation:

Close your eyes and direct your attention to your breathing. Allow it to become slower and more rhythmic. Notice the abdomen expand as you inhale and contract as you exhale. Take a few moments to concentrate on your breathing.

Now that you are relaxing, clench the right fist. Clench tighter and tighter, noting the tension in the right hand. Hold the tension for about five seconds, then quickly release the tension and allow the right hand to relax. Note the sense of relaxation in the right hand and how this differs from tension. Repeat the tense-relax procedure with the left hand.

Next tense your right arm. Hold the tension for about five seconds and note the feeling of tension. Now release the tension quickly and focus on the sense of relaxation in the right arm. Be aware of how this sensation differs from tension. Repeat the tense-relax procedure with the left arm.

Focus now on your right leg. Tense the leg, hold the tension, then release it quickly. Concentrate on the feeling of relaxation in the right leg and how this differs from tension. Repeat the tense-relax procedure with the left leg.

The muscles are becoming more relaxed and the breathing slower and more rhythmic. If your mind drifts to distracting thoughts, return your attention to the feeling of relaxation in the body.

Now tense the abdomen. Hold the tension for five seconds and notice the feeling of tension in the abdomen. Quickly release the tension and be aware of how relaxation differs from tension. Repeat the tense-relax procedure for the back muscles, noting the difference between tension and relaxation. Tense and relax the neck and shoulder muscles, then the jaw, cheeks, eyes, and forehead, concentrating on the difference between tension and relaxation in these muscles.

Spend a few moments feeling relaxation in your body. Note the slower rhythmic breathing pattern. Disregard any distracting thoughts. You are becoming more deeply relaxed, calm, and peaceful. Take a few moments to concentrate on relaxation.

Finally, at your own pace, take a slow, deep breath and slowly open your eyes.

Notes

Introduction: From Whence We Came

1. According to Dr. Brent Hafen and his colleagues, authors of *Mind/ Body Health* (Boston: Allyn and Bacon, 1996), the American Institute of Stress estimates that 75 to 90 percent of all visits to health care providers result from stress-related complaints, while the American Academy of Family Physicians estimates that two-thirds of all office visits to family doctors are prompted by stress-related symptoms.
2. Hafen, et al., *Mind/Body Health*.
3. Ibid.
4. See Paul Recer, "Four Million Abuse Prescription Drugs, U.S. Aide Says," *The Boston Globe*, April 11, 2001.
5. Although I refer to the modern mind as the Thinking Mind, we will see that ancestral man also had the capacity for complex thought. Nevertheless, the modern Thinking Mind is distinct from the Ancestral Mind in that the TM is characterized by rationalism and self-reflective thought.

Chapter 1: The Tyranny of the Thinking Mind

1. Mihaly Csikszentmihalyi, *Flow: Toward a Psychology of Optimal Experience* (New York: HarperPerennial, 1990).
2. Julian Jaynes, *The Origins of Consciousness in the Breakdown of the Bicameral Mind* (Boston: Houghton Mifflin, 1976).
3. Henri Frankfort et al., *Before Philosophy: The Intellectual Adventure of Ancient Man* (Baltimore: Penguin Books, 1967).
4. Thomas Moore, *The Re-Enchantment of Everyday Life* (New York: HarperPerennial, 1996).
5. For a highly readable discussion of the perception of time, work, and leisure in preindustrial society, see the chapter entitled "Civilization and Its Displeasures" in Robert Ornstein and David Sobel, *Healthy Pleasures* (Reading, Mass.: Addison-Wesley, 1989).
6. Juliet Schorr, *The Overspent American* (New York: Basic Books, 1998).
7. Dr. Martin Clarkberg and colleagues presented their findings of a study on the work-family time squeeze at the 1999 American Asso-

ciation for the Advancement of Science meeting in Anaheim, California. The study was based on surveys of 4,554 married couples questioned in 1988 and 1994 as part of the National Study of Families and Households.

8. See the *Boston Globe* interview with T. Berry Brazelton, October 29, 2000.

9. See David Myers and Ed Diener, "Who Is Happy?," *Psychological Science* 6 (1995), 10–19.

10. Ibid.

11. See Tim Kasser and Richard Ryan, "A Dark Side of the American Dream: Correlates of Financial Success as a Central Life Aspiration," *Journal of Personality and Social Psychology* 65 (1993): 410–22.

12. Myers and Diener, "Who Is Happy?"

13. David Myers, "The Funds, Friends, and Faith of Happy People," *American Psychologist* 55 (2000): 56–57.

14. Schwartz, Barry, "Self-determination: The Tyranny of Freedom," *American Psychologist* 55 (2000): 79–88. Schwartz raises an interesting question: Why is it that, for the first time in human history, large numbers of people can live unconstrained lives with high levels of choice, yet there is a concurrent explosion in depression rates? Schwartz and other psychologists believe that part of the answer is that excessive choice can become a major burden instead of a blessing, requiring a rapidly changing worldview and adaptation that is highly stressful. The burden of responsibility for making choices can result in a kind of psychological tyranny—an excess of freedom that may lead to insecurity and regrets. When the burden of choice becomes too heavy, we may become dissatisfied and depressed.

15. For a thought-provoking discussion of the stress of modernization, see the chapter entitled "Clinical Applications" by Robert Woolfolk and Paul Lehrer in their *Principles and Practice of Stress Management* (New York: The Guilford Press, 1984).

16. See Adam Pertman, "Information Overload," *The Boston Globe*, February 11, 2001.

17. See David Armstrong, "Exaggerating Our Lives," *The Boston Globe*, August 30, 1998.

18. Ibid.

19. B. S. Centerwall, "Television and Violence: The Scale of the Problem

and Where to Go from Here," *Journal of the American Medical Association* 10 (1992): 3059–63. Centerwall showed that homicide rates were correlated with the introduction of television. For example, he found that, as the first generation of American and Canadian children exposed to television reached crime-committing age, the homicide rate climbed steeply and eventually doubled. In South Africa, where television was banned until 1975, the low rate of white homicide deaths more than doubled when white children reached crime-committing age. Centerwall concluded that long-term childhood exposure to television violence is a causal factor in approximately one-half the homicides committed in the United States, or about 10,000 homicides annually, and that exposure to television violence is a major cause of rapes, assaults, and other interpersonal violence. Based on these data, Centerwall hypothesized that, if television technology had never been developed, there would be today 10,000 fewer homicides annually in the United States, 70,000 fewer rapes, and 700,000 fewer assaults.

20. See S. Villani, "Impact of Media on Children and Adolescents: A 10-Year Review of the Research," *Journal of the American Academy of Child and Adolescent Psychiatry* 40 (2001): 392–401.

21. Ibid.

22. Ibid.

23. Ibid.

24. David Buss, "The Evolution of Happiness," *American Psychologist* 55 (2000): 15–20.

25. Ibid.

26. Ibid. Scientists have also postulated that the media may play a fundamental role in the increased prevalence of depression in modern life. Depression rates are not only increasing, they are higher in developed countries. In attempting to answer why rates of depression are climbing despite the abundance of comforts and technological solutions to so many ancient problems, R. M. Nesse and G. C. Williams (*Why We Get Sick* [New York: New York Times Books, 1994]) offer this hypothesis: "Mass communications, especially television and movies, effectively make us all one competitive group even as they destroy our more intimate networks. In the ancestral environment you would have had a good chance at being the best at some-

thing. Even if you were not the best, your group would likely value your skills. Now we all compete with those who are the best in the world. Watching these successful people on television arouses envy. Envy probably was useful for our ancestors to strive for what others could obtain. Now few of us can attain the fantasy lives we see on television." Thus, according to Nesse and Williams, the increase in depression in modern life stems from self-perceived failures resulting from inaccurate comparisons between people's lives and the glamorous lives depicted in the media. Our own spouses, parents, and children can seem profoundly inadequate by comparison. Consequently, we are dissatisfied with them and ourselves.

27. Erich Fromm, *Zen Buddhism and Psychoanalysis* (New York: Harper and Row, 1960).

Chapter 2: What *Is* This Ancestral Mind?

1. The importance of the Mammalian Brain as the anatomical basis for the experience of emotion was first described in the 1930s by James Papez (see J. W. Papez, "A Proposed Mechanism of Emotion," *Archives of Neurology and Psychiatry* 79 [1937]: 217–24) and then more fully by Paul MacLean, who initially used the term "limbic system" to describe this part of the brain (see P. D. MacLean, "Psychosomatic Disease and the 'Visceral Brain': Recent Developments Bearing on the Papez Theory of Emotion," *Psychosomatic Medicine* 11 [1949]: 338–53.) The terms *limbic system* and *mammalian brain* are often used interchangeably.

   The limbic system is complex and, because there are still many inconsistent findings concerning it, there is no current consensus on its structure and precise function. It is more common to refer to limbic "structures" than limbic "system," for these structures probably don't constitute a distinct system. In fact, Joseph LeDoux has recently called for the abolishment of the term *limbic system*, as he believes that there is not one but multiple emotional systems in the brain. See Joseph LeDoux, *The Emotional Brain* (New York: Simon & Schuster, 1996).

2. After being discovered approximately fifty years ago, the concept of the reticular formation fell into disuse until more recently, when neuroscientists used more modern brain imaging and staining tech-

niques to define this system more clearly and substantiate its importance in brain functioning. See M. Steriade, "Arousal: Revisiting the
Reticular Activating System," *Science* 272 (1996): 225–26.

3. According to J. D. French, one of the seminal researchers on the reticular formation, it not only awakens the brain to consciousness but
contributes in an important way to the highest mental processes like
introspection. See J. D. French, "The Reticular Formation," *Scientific
American*, May 1957, 2–8.

4. The thalamus performs its gating function by enhancing or inhibiting sensory messages. Some neuroscientists believe that, like the
thalamus, the RF also plays an important role in selective attention
by opening or closing the flow of sensory information that streams
into the brain stem. For this reason, they view the thalamus and RF
as being inextricably linked in mediating attention and consciousness. See J. Newman, "Thalamic Contributions to Attention and Consciousness," *Consciousness and Cognition* 4 (1995): 172–93.

5. One reason why we know so much about the amygdala's role in negative emotions and less about its role in positive emotions is that negative emotions such as fear are easier to study in animals. Although
the weight of the evidence highlights the role of human amygdala
activation in response to fear-producing stimuli, Paul Whalen's research has demonstrated amygdala activation in response to subtle
emotional stimuli such as photographs of facial expressions, which
are not clearly related to fear. According to Whalen, amygdala activation in response to faces may be related to the ambiguity of facial
expressions. He believes that compartmentalizing amygdala functioning as fear-related does not allow for appreciation of "subtler"
emotional experiences that speak to a greater portion of our daily experience. He argues for a broader theory of the amygdala to include
fear, vigilance, and ambiguity. See P. J. Whalen, "Fear, Vigilance, and
Ambiguity: Initial Neuroimaging Studies of the Human Amygdala,"
Current Directions in Psychological Science 7 (1998): 177–86.

6. The eminent Harvard physiologist Walter Cannon was the first to
describe the neurophysiological correlates of the stress response:
"Respiration deepens, arterial pressure rises, the blood is shifted
away from the stomach and intestines to the heart and the central
nervous system and muscles, the spleen contracts and discharges its

content of concentrated corpuscles, and adrenalin is excreted." Cannon called the stress response the "emergency reaction," which he conceptualized as a specific physiological response of the body that accompanies any state in which physical energy must be exerted. He believed that blood flow is redistributed to areas of the muscles and bodily organs that are active during emergencies. See Walter Cannon, *The Wisdom of the Body* (New York: W. W. Norton, 1932). Then, working in the 1950s, the Swiss Nobel laureate Walter Hess demonstrated that electrical stimulation of the hypothalamus produced the physiological changes that Cannon described twenty years earlier. See Walter Hess, *Functional Organization of the Diencephalon* (New York: Grune and Stratton, 1957).

7. See Andrew Newberg, *Why God Won't Go Away* (New York: Ballantine, 2001). Newberg points out that repetitive sensory stimulation can be slow, as in meditation, or fast as in dancing, sexual activity, or the frenzy of a shaman. In fast, repetitive stimulation, arousal systems are driven at a maximum so that a massive quiescent rebound ensues at the same time that the orientation cortex is receiving repetitive sensory input. Although he believes that the brain processes by which slow and fast repetitive activities alter the functioning of the orientation cortex are slightly different, the resulting state is the same—one of relaxation, pleasurable calm, and dissolution of the boundaries of the self. The degree to which the orientation cortex is deprived of sensory input determines the intensity of changes in self-consciousness: more profound alterations in sensory input result in more profound altered states of awareness such as awe, rapture, or mystical states, while minor changes lead to states of flow or absorption. Newberg suggests that ritualized behaviors such as courtship and holidays also affect the self by aiming to define the individual as part of a larger group or cause. Some animals even exhibit ritualized behaviors that transcend self-protective behaviors.

8. See Daniel Goleman, *Emotional Intelligence* (New York: Bantam Books, 1995).

Chapter 3: Bypassing the Thinking Mind

1. Many of these studies were conducted in the late 1800s and early 1900s when consciousness was a central topic in psychology. For ex-

ample, studies show that the following types of information can be perceived unconsciously: visual characters such as letters or digits; small differences in the weight of two objects; geometric figures and orientations of lines such as vertical or diagonal; the meaning of words. Taken together, these results show that a considerable amount of information is perceived even when observers do not report any awareness of perceiving. For a recent review of studies demonstrating unconscious perception of cognitive information, see P. Merikle and M. Daneman, "Conscious vs. Unconscious Perception," in *The New Cognitive Neurosciences*, ed. M. Gazzaniga (Cambridge, Mass.: MIT Press, 2000).

2. For an excellent and concise summary of subliminal perception research, see the chapter entitled "Blood, Sweat, and Tears" in Joseph LeDoux, *The Emotional Brain* (New York: Simon & Schuster, 1996).

3. Ibid.

4. See R. Zajonc, "Feeling and Thinking: Preferences Need No Inferences," *American Psychologist* 35 (1980): 151–75. Zajonc presents persuasive evidence for the role of unconscious processing in emotions. He argues on the basis of numerous experiments involving subliminal presentation of emotional stimuli that simple emotional reactions, called preferences, could be formed without any conscious registration of emotional stimuli.

5. LeDoux provides a fascinating account of his research on the amygdala in *The Emotional Brain.*

6. See Paul Whalen et al., "Masked Presentations of Emotional Facial Expressions Modulate Amygdala Activity Without Explicit Knowledge," *Journal of Neuroscience* 18 (1998): 411–18.

Chapter 4: The Power of Positive Emotions

1. For an excellent summary of scientific thinking about emotions, see Paul Ekman and Richard Davidson, eds., *The Nature of Emotion* (New York: Oxford Press, 1994).

2. Harvard Medical School physician Jerome Groopman, a leading clinician and researcher in cancer and AIDS, explored the ways in which intuition informs medical decisions in his book *Second Opinions: Stories of Intuition and Choice in the Changing World of Medicine* (New York: Penguin, 2001). He uses extensive case studies to show

that medical outcomes hinge as much on a doctor's gut feelings as on his expertise. It is a combination of a doctor's knowledge and his intuition that builds a "medical compass" that guides diagnoses and the safest, least traumatic treatment for people who are in advanced stages of illness. He argues that medicine is a balance between the "sixth sense" of intuition and medical knowledge.

3. Emotions allow us to unconsciously and quickly deal with situations that affect our well-being in ways that have been repeatedly successful through our individual past and our ancestors' distant evolutionary past. From an evolutionary perspective, rapid, unconscious emotional responses are more advantageous than the TM's slower conscious reflection. The primary function of the brain is to ensure survival, not by calculating and planning but by quickly mobilizing an organism to respond to events that are crucial to its well-being. Because thinking, reflection, and planning are, in evolutionary terms, secondary to the wisdom of the past that is needed for survival, the brain defers to the intuitive intelligence of the AM, not the rational intelligence of the TM, in matters of importance to our well-being. The prominent role that evolution has given the AM in well-being is so strong that it is wired into the brain; the neural connections from the AM to the TM are stronger than connections from the TM to the AM, allowing the AM to exert influence over the TM in matters of importance.

The wisdom of emotions is imprinted into the AM as unconscious memories that carry the genetic knowledge of our distant ancestral past. The presence of these eternal genetic memories, held in common throughout man's evolutionary history, is evidence of the existence of archetypal ancestral memories. Some emotions researchers believe that it is not unreasonable to suppose that many myths and legends, passed down through early oral traditions, reflected the existence of these archetypal memories in the brain. As emotions researcher Jaak Panksepp puts it, "Traditions of folklore may reflect the ease with which the basic emotional feelings are capable of being linked to world events (to specific scripts) to provide powerful symbolic representations of our evolutionary heritage." See Jaak Panksepp, "Subjectivity May Have Evolved in the Brain as a Simple, Value-Coding Process That Promotes the Learning of New Behaviors," in Ekman and Davidson, *The Nature of Emotion*.

4. Joseph LeDoux, *The Emotional Brain* (New York: Simon & Schuster, 1996).

5. Antonio Damasio, *Descartes' Error* (New York: Putnam, 1994).

6. See Paul Ekman, "Moods, Emotions, and Traits," in Ekman and Davidson, *The Nature of Emotion*.

7. Over the past fifty years, psychology has focused primarily on mental illness, fear, and aggression rather than on health and positive emotions, traits, and behaviors. Since 1887, there have been over 135,000 articles published on anger, anxiety, and depression compared to less than 3,000 articles on happiness. Most of what we know about emotions pertains to negative emotions like fear because, from a scientific and evolutionary perspective, fear is perhaps the most important emotion—an organism's principal concern is surviving danger. Negative emotions are also important for understanding and treating psychological disorders such as depression or post-traumatic stress disorder. As a result, psychology has, according to Drs. Martin Seligman and Mihaly Csikszentmihalyi, developed a distorted view of what normal and exceptional human experience is like. Seligman and Csikszentmihalyi assert that psychologists have scant knowledge of what makes life worth living. They believe that psychologists know a lot about adversity and mental illness but very little about the other fundamental missions of psychology—making the lives of all people better and understanding how people flourish under more normal conditions—in large part because psychology has been a science of healing pathology while neglecting the fulfilled individual. They point to two events—the founding of the Veterans Administration and the National Institutes of Mental Health (NIMH)—as providing the major impetus for the study of pathology. The VA employed thousands of psychologists to treat mental illness, while the NIMH, which was based on the disease model, showed psychologists that they could get grants if their research was about pathology (Seligman and Csikszentmihalyi suggest that the NIMH should be more appropriately renamed the National Institute of Mental Illness). For a review of their and others' arguments for a positive

psychology that includes the study of strength and virtue, see the entire special issue of the *American Psychologist* dedicated to positive psychology (vol. 55, January 2000), edited by Seligman and Csikszentmihalyi.

8. Psychologist Richard Davidson believes that most basic emotions involve approach or withdrawal: positive emotions involve approach-related behavior while negative emotions involve withdrawal-related tendencies. However, there is no definitive evidence that all positive emotions involve approach, and some emotions such as anger can involve either approach or withdrawal. Robert Thayer, a psychologist at California State University, Long Beach, and a leading authority on emotions, believes that positive and negative emotions can also be explained by variations in two continuums: "tense-tired" and "calm-energy." He holds that when we are tense and tired, we experience negative feelings, moods, and thoughts; when we are calm and energetic, positive feelings result. Robert Thayer, *The Origin of Everyday Moods* (New York: Oxford Press, 1996).

9. Researcher Paul Ekman proposed a theory of emotions that consists of six basic emotions: surprise, happiness, anger, fear, disgust, and sadness. This theory is based on his discovery that specific facial expressions are recognized in cultures worldwide, including preliterates. Another researcher, Sylvan Tompkins, proposed the existence of eight basic emotions: surprise, interest, joy, rage, fear, disgust, shame, and anguish. Most emotion theorists assume that there are secondary emotions that represent combinations of basic emotions. See Ekman and Davidson, *The Nature of Emotion.*

10. See R. J. Davidson, R. E. Wheeler, and R. C. Doss, "Individual Differences in Anterior Brain Asymmetry and Fundamental Dimensions of Emotion," *Journal of Personality and Social Psychology* 62 (1992): 676–87. Decreased activation in the left prefrontal lobes for negative emotions have also been observed by Michael Gemar and his colleagues (M. C. Gemar et al., "Effects of Self-Generated Sad Mood on Regional Cerebral Activity: A PET Study in Normal Subjects," *Depression* 4 [1996]: 81–88).

11. See J. B. Henriques and R. J. Davidson, "Left Frontal Hypoactivation in Depression," *Journal of Abnormal Psychology* 100 (1991): 535–45. Other investigators, using PET (positron emmission tomography, a

brain-imaging technique), have also found reduced activation in the left prefrontal lobes in depressed patients. See L. R. Baxter et al., "Cerebral Metabolic Rates for Glucose in Mood Disorders: Studies with Positron Emission Tomography and Fluorodeoxyglucose F 18," *Archives of General Psychiatry* 42 (1985): 243–50.

12. See G. D. Jacobs and D. Snyder, "Frontal Brain Asymmetry Predicts Affective Style in Men," *Behavioral Neuroscience* 110 (1996): 3–6.

13. Interestingly, intense emotions like ecstasy are not as important to well-being because they are rarer in daily life. This finding may also be explained by the fact that intense emotions are short-lived. Over time, the immediate emotional response to significant life events fades. Thus, both the pleasure and the agony of emotional events gives way to a return of more normal emotions. See E. Diener and R. J. Larsen, "The Experience of Emotional Well Being," in *Handbook of Emotions*, ed. M. Lewis and J. M. Haviland (New York: Guilford Press, 1993).

14. Due to a paucity of studies on the neuroanatomical correlates of positive emotions in humans, we know little about the specific circuitry of positive emotions and their subjective states. While some studies have reported decreases in activity of brain areas involved in emotions during happiness (see M. S. George et al., "Brain Activity During Transient Sadness and Happiness in Healthy Women," *American Journal of Psychiatry* 152 [1995]: 341–51), others have reported increased activity (R. D. Lane et al., "Neuroanatomical Correlates of Happiness, Sadness, and Disgust," *American Journal of Psychiatry* 154 [1997]: 926–33).

In many of these studies, happiness is studied by having subjects remember happy events and trying to make themselves feel that way again; this may be quite different from the actual experience of happiness in real life. People are much less likely to express the strong emotions of real life in a laboratory with an experimenter present, in part because laboratory stimuli designed to elicit emotions are not the same as the emotional stimuli of real life. It is also possible that some subjects may fail to induce happiness when instructed to do so, may not be generating the same emotion as other subjects, or may be generating different intensities of an emotion compared to other subjects (happiness can range from contentment to euphoria).

Also, there is no agreement on what represents an appropriate baseline or comparison condition in these studies (comparison conditions typically involve instructions to think about "neutral" memories, but these may, in fact, not be neutral). The baseline problem is complicated by the fact that, during a resting baseline, the brain exerts a significant amount of energy to inhibit the normal barrage of sensory input into the brain (in other words, the default baseline of the conscious resting brain is activation). States of focused attention are characterized by a deactivation of brain structures such as the cerebral cortex due to the reduction in energy normally required to inhibit sensory input. Because many baseline conditions in brain imaging studies use focused attention on neutral images or stimuli, any change from a deactivated baseline may involve activation. For a discussion of these issues, see M. Rachael, "The Neural Correlates of Consciousness: An Analysis of Cognitive Skill Learning," in *The New Cognitive Neurosciences*, ed. M. Gazzaniga (Cambridge, Mass.: MIT Press, 2000).

These methodological problems may explain the lack of consistent findings on the anatomical correlates of positive emotions across studies and why the specific correlates of positive emotions remain unclear.

Although mind/body techniques like the relaxation response (RR) induce positive emotional states, no well-controlled studies have used PET or fMRI to determine the effects of the RR on the brain. However, EEG and physiological studies on the RR suggest that it deactivates the brain. Also, studies on sleep demonstrate that states of lower arousal deactivate the brain. As we will see in chapter nine, stage 1 sleep is similar to the RR. Therefore, it is likely that the RR deactivates brain structures in the AM. Also, the fact that focused attention deactivates brain structures suggests that mind/body techniques like the RR that induce pleasant emotional states by focusing attention on a repetitive mental stimulus would deactivate brain structures. Even in ecstatic emotional states, there may be a deactivation in the brain if the ecstatic state is the result of focused attention.

15. Richard Davidson and Nathan Fox, *Journal of Abnormal Psychology* 98 (1989): 127–31.

16. These findings come from Jerome Kagan's research on temperament in infants and toddlers at Harvard University. For a very interesting and readable summary of Kagan's research, see the chapter entitled "Temperament Is Not Destiny" in Daniel Goleman, *Emotional Intelligence* (New York: Bantam, 1995).

17. See J. Leserman et al., "Sexual and Physical Abuse History in a Gastroenterology Practice: How Types of Abuse Impact Health Status," *Psychosomatic Medicine* 58 (1996): 4–15.

18. Scientists like Richard Davidson call this model, in which patterns of brain activity present at birth interact with environmental stressors, a "stress-diathesis model." Brain activity patterns represent a diathesis that increases our vulnerability to emotional styles and disorders like depression but this vulnerability may not be expressed without the requisite environmental stress. The model holds that brain patterns present at birth are, by themselves, not enough to produce a specific emotional style; the environment must also play a role.

19. See the chapter "Temperament Is Not Destiny" in Goleman, *Emotional Intelligence*.

20. Meta-analytic studies on the efficacy of pharmacological and cognitive-behavioral therapies for psychiatric problems such as depression, panic disorder, and phobias reveal that cognitive-behavioral therapies are at least as effective as, and in some cases more effective than, medication. Some studies have shown that the most effective intervention for these disorders involves a combination of cognitive-behavioral and pharmacological therapies.

21. S. Epstein, "Integration of the Cognitive and Psychodynamic Unconscious," *American Psychologist* 44 (1994): 710–17.

22. In *Emotional Intelligence*, Goleman cites international data that suggest a modern epidemic of depression. Each successive generation worldwide since 1900 has lived with a greater risk of depression than their parents. Depression is also beginning at earlier ages as evidenced by the emergence of childhood depression, which was unrecognized until recently. In *Why We Get Sick* (New York: New York Times Books, 1994), R. Nesse and C. Williams cite data from 39,000 people in nine different countries carried out in five diverse areas of the world showing that young people in each country are far more

likely than their elders to have experienced an episode of major depression. Furthermore, the rates were higher in societies with higher degrees of economic development.

Chapter 5: The Ancestral Way of Life

1. R. Leakey and R. Lewin, *Origins* (New York: E. P. Dutton, 1977).
2. Ibid.
3. Ibid. As Leakey points out, the idea of prehistoric man eking out an existence from nature without technology seems a daunting task. In reality, for at least several million years our ancestors followed a technologically simple but highly successful way of life that allowed man to emerge as the dominant species on earth. One reason for this success was the making and using of tools, not weapons, which allowed for scavenging and obtaining food. Another was that they pursued a hunting and gathering lifestyle that demanded a high degree of cooperation and interdependence not displayed by other primates, which allowed them to exploit resources. Given that this lifestyle was part of our evolution over millions of years, the hunter-gatherer existence was a potent force, and is an indelible genetic memory, in what makes us human.
4. Ibid.
5. Ibid.
6. Like modern man, ancestral man was capable of violence and homicide. According to L. H. Kelley, author of *War Before Civilization* (New York: Oxford University Press, 1996), there is no consistent evidence of homicide in the archaeological record before about 10,000 years ago (many of the traumas found on early hominid skeletons have been proved by subsequent investigation to have had nonhomicidal causes and cannot be distinguished from accidental traumas of a similar character such as puncture created by leopard canines). However, several rare burials of early modern humans in Europe, dating from 34,000 to 24,000 years ago, show evidence of violent death. Human skeletons found in Egypt, dating about 12,000 to 14,000 years ago, show that warfare was practiced and was brutal. (Over 40 percent of the fifty-nine skeletons of men, women, and children had stone projectiles associated with or buried in them.) Archaeological evidence appears to indicate that warfare is consistently docu-

mented in the archaeological record of the past 5,000 to 10,000 years in many regions.

7. Ibid.

8. According to anthropologist Don Johansen, who discovered Lucy, the oldest human fossil find at 3.5 million years old, Cro-Magnons were not the lurching subhumans of popular belief; they were men like us. They presumably could think complex thoughts, exhibited the beginnings of religious beliefs, used tools routinely, and had an elaborate culture. Neanderthal man was heavier boned and had more primitive facial features, but he was also a man. According to Johansen: "One hears about putting him in a business suit and turning him loose in the subway. It is true; one could do it and he would never be noticed. Could he make change at the subway booth and recognize a token? He certainly could. He could do many things more complicated than that. He was doing them over much of Europe, Africa, and Asia as long as sixty or a hundred thousand years ago." D. Johansen and M. Edey, *Lucy: The Beginnings of Humankind* (New York: Simon & Schuster, 1981).

9. See D. Chandler, "Ancient Note: Music as a Bridge Between the Species," *The Boston Globe,* January 5, 2001.

10. Although we should not imagine that the !Kung represent the exact way our ancestors lived, the !Kung exhibit the same patterns of behavior as our distant prehistoric ancestors. And although the !Kung practiced a cooperative lifestyle, some scientists believe that they were not entirely peaceful. These scientists cite evidence that the !Kung conducted small-scale raids and engaged in feuds and homicides between other intruding bands and herders. In many societies, members can be extremely cooperative toward one another and yet be very aggressive and violent toward outsiders, particularly when resources deteriorate during disasters or territorial expansion. See L. H. Kelley, *War Before Civilization*.

11. Leakey and Lewin, *Origins*.

12. The Sufis are but one example of esoteric traditions whose teachings focus on people who are too preoccupied to hear what is being said, who do not see what is in front of them because of their internal monologue. The Sufis emphasize the constantly changing biases that constitute normal awareness and how our needs sometimes block us from a

more comprehensive perception of reality. They characterize ordinary TM consciousness as a state of "blindness." See Robert Ornstein, *The Psychology of Consciousness* (San Francisco: W. H. Freeman, 1972).

13. Mihaly Csikszentmihalyi, *Flow: The Psychology of Optimal Experience* (New York: Harper and Row, 1991).

14. See Abraham Maslow, *Religions, Values, and Peak Experiences* (New York: Viking, 1971), and *Toward a Psychology of Being* (New York: Van Nostrand, 1968).

15. In its most developed, and unusual, form, the ancestral state of mind has been called the mystical experience by esoteric traditions from ancient Hindu to contemporary European, from the Bible to the Koran. This state of consciousness is qualitatively different from our normal TM-based self-consciousness and is called mystical because it is beyond the scope of language, rationality, and the TM. Individual consciousness is felt to merge with something much vaster than our self, which is felt to be "universal." As mystics of many religions have taught, the feeling of union can give a sense of eternal fulfillment that makes life vibrant.

16. The poet-scientist Edmund Carpenter described the mystical state of consciousness this way three quarters of a century ago:

Of all the hard facts of science, I know of none more solid and fundamental than the fact that if you inhibit thought you come at length to a region of consciousness below or behind thought, and different from ordinary thought in its nature and character—a consciousness of quasi-universal quality, and a realization of an altogether vaster self than that to which we are accustomed. And since the ordinary consciousness, with which we are concerned in ordinary life, is before all things founded on the local little self, and is in fact self-conscious in the little local sense, it follows that to pass out of that is to die to the ordinary self and the ordinary world. It is to die in the ordinary sense but, in another sense, it is to wake up and find that the "I," one's real, most intimate self pervades the universe and all other beings—that the mountains and the sea and stars are part of one's body and that one's soul is in touch with the soul of all creatures. So great, so splendid is this experience that it may be said that all minor doubts and questions fall away in face of it; and certain it is that in thousands and thousands of cases the fact of its having come even once to a man has completely revolutionized his subsequent life and outlook on the world. (Quoted in J. White, *The Highest State of Consciousness* [New York: Doubleday, 1972]).

17.   Awe has been so consistently associated with powerful positive emotional states like flow, the peak experience, and the mystical experience that it is becoming increasingly studied by emotions researchers. See Gareth Cook, "Seeing How the Spirit Moves Us," *The Boston Globe*, December 6, 2000.

18.   See A. Luks, *The Healing Power of Doing Good: The Health and Spiritual Benefits of Helping Others* (New York: Fawcett Columbine, 1991).

19.   Cook, "Seeing How the Spirit Moves Us."

20.   The desire to understand the conditions under which organisms help one another, cooperate, share, and even sacrifice themselves has generated some of the most important recent advances in evolutionary biology. In the past, biologists believed that it would be difficult for altruism to evolve and that organisms were designed to be selfish since selfishness would increase the likelihood of survival. However, evolutionary biology has discovered that altruism and cooperation are, in fact, as natural as selfishness. Many organisms will endanger themselves to aid another organism (people will put themselves at great peril to help their children or others, chimpanzees help one another in fights, vampire bats feed the blood that they have collected to other bats in need, and squirrels emit warning sounds to other squirrels about the presence of a predator even though the warning sound can draw the predator's attention to the squirrel). Since endangering one's life would appear to decrease the likelihood of successful reproduction, why would such a design be selected for in the process of evolution?

The answer is that altruistic behaviors increase the fitness, or reproductive success, of the organisms' kin, relatives, or nonrelatives, thus increasing the fitness of the species. Also, the delivery of benefits to nonrelatives increases the probability that benefits will be delivered in return. For example, pooling of food resources among nonrelatives reduces the likelihood of starvation during periods of famine. Food sharing can therefore be viewed as a type of social obligation; by accepting food, others obligate themselves to reciprocate in the future. Because food acquisition was so integral to ancestral life (our ancestors spent more waking hours in searching for food than in any other activity), behaviors such as food sharing probably had a critical impact on ancestral life and evolution. Hunting, a high-variance activity in which luck plays an important role, may have been a particularly important force in the evolution of cooperation and altruism; an individual might be successful one week but not the next. Also, because food spoiled rapidly, meat from animals came in quantities that far exceeded what a single hunter could consume. For these reasons, hunting favored sharing. See L. Cosmides and J. Tooby, "Cognitive Adaptations for Social Exchange," in *The Adapted Mind: Evolutionary Psychology and the Generation of Culture*, ed. J. Barkow, L. Cosmides, and J. Tooby (New York: Oxford University Press, 1992).

Altruistic behaviors could also serve as a foundation for morality that leads to reciprocal altruism, in which group members trade favors. Committing such acts would have significant value in bolstering survival and may lead to the capacity for sympathy and trust. From this perspective, morality may be a core part of the AM that is driven by instinct and emotion. See J. Barkow, "Happiness in Evolutionary Perspective," in *Uniting Psychology and Biology*, ed. N. Segal, G. Weisfeld, and C. Weisfeld (Washington, D.C.: American Psychological Association, 1997).

Cooperation can also occur across different species or when friendship is absent. The live-and-let-live system that emerged in the trench warfare of World War I demonstrates that cooperation can develop between antagonists. This system was endemic in trench warfare; it flourished despite the efforts of senior officers to stop it, the emotions involved in combat, and the military logic of kill or be killed. See R. Axelrod, *The Evolution of Cooperation* (New York: Basic Books, 1984).

21. Cook, "Seeing How the Spirit Moves Us."

Chapter 6: Taming Toxic Thoughts

1. According to Herbert Benson, Eileen Stuart, and the staff of the Mind/Body Medical Institute at Harvard Medical School, authors of

The Wellness Book (New York: Fireside, 1992), our NATs are typically the result of the following common cognitive distortions:

- *All-or-nothing thinking.* We tend to evaluate stressful situations in black-or-white categories, as in the straight-A student who received his first B and concluded "Now I am a failure." This type of thinking is often the basis of perfectionism.
- *Overgeneralization.* We see a single negative event as part of a consistent pattern of defeat.
- *Mental filter.* We pick out a negative detail from a situation and dwell on that exclusively, thus perceiving the whole situation as negative.
- *Discounting the positive.* This involves the tendency to transform positive or neutral experiences into negative ones. An example is receiving a compliment and thinking "They don't really mean it."
- *Jumping to conclusions.* This involves concluding the worst when it is not justified by the facts. This can involve mind reading, in which you assume someone is thinking negatively about you when they are not, or fortune telling, in which you anticipate that something will turn out negatively in the future.
- *Magnification.* We exaggerate the significance of a negative event.
- *"Should" statements.* These make us feel pressured and can result in frustration when they are not met.
- *Personalization and blame.* You assume responsibility for a negative event when there is no basis for doing so; or, blame others for negative events with no basis.

2. See David Burns, *The Feeling Good Handbook* (New York: Plume, 1989). For an in-depth "workbook" description of cognitive restructuring, see D. Greenberger and C. Padesky, *Mind over Mood: A Cognitive Therapy Treatment Manual for Clients* (New York: The Guilford Press, 1995).

3. Cognitive restructuring therapy was developed by Aaron Beck at the University of Pennsylvania in the early 1960s as a structured, short-term, present-oriented therapy for depression. Beck proposed that distorted or dysfunctional thinking, which influences the patient's mood and behavior, is common to all emotional disorders. Various forms of CR therapy have been developed by other clinicians, including Albert Ellis's rational-emotive therapy, Don Mei-

chenbaum's cognitive-behavioral modification, and David Burns's CR therapy.

4. For a concise review of the efficacy of cognitive restructuring for depression, see Aaron Beck, "Cognitive Therapy: Past, Present, and Future," *Journal of Consulting and Clinical Psychology* 61 (1993): 194–98.

5. For an overall review of the various disorders to which cognitive restructuring has been successfully applied, see Judy Beck, *Cognitive Therapy: Basics and Beyond* (New York: The Guilford Press, 1995).

6. Ibid.

Chapter 7: The Power of Stress-Reducing Attitudes and Beliefs

1. Whereas the self-monologue is more cognitive in nature, attitudes and beliefs have inherent emotional and motivational components. Indeed, some psychologists conceptualize attitudes and beliefs like optimism or altruism as moods or feelings, for there is an emotional flavor to them: We *feel* optimistic or altruistic and these feelings motivate us. See C. Peterson, "The Future of Optimism," *American Psychologist* 55 (2000): 44–55.

   Although attitudes and beliefs can be considered cousins of the internal monologue, they differ in other important ways. First, attitudes and beliefs are more unconscious; they are akin to filters through which we view the world. The filters influence our mental monologue but are not as conscious. For example, consider a negative automatic thought (NAT) such as "I am not looking forward to changing jobs." This conscious NAT may be the result of a more unconscious underlying belief such as "change is threatening." We will see that underlying beliefs such as this can provoke stressful thoughts and emotions. Second, attitudes and beliefs are more ancient than the internal monologue. Although some scholars believe that self-conscious thought is likely a recent development in the evolution of the mind, archaeological evidence suggests that ancestral man exhibited the capacity for beliefs tens of thousands of years earlier.

2. Richard Davidson presented his research in a lecture at the 2000 Wisconsin Symposium on Emotion entitled "The Neurobiology of Positive Emotion."

3. Lionel Tiger, *Optimism: The Biology of Hope* (New York: Simon & Schuster, 1979).

4. M. E. P. Seligman, *Learned Optimism* (New York: Knopf, 1991).

5. Ibid.

6. See T. Monmaney, "'Have a Nice Day' May Translate to a Nicer Life," *The Boston Globe*, January 4, 2000.

7. C. Peterson and L. M. Bossio, *Health and Optimism* (New York: The Free Press, 1991). See also Seligman, *Learned Optimism*.

8. Peterson, "The Future of Optimism."

9. Seligman, *Learned Optimism*.

10. C. Peterson et al., *Learned Helplessness: A Theory for the Age of Personal Control* (New York: Oxford University Press, 1993).

11. Seligman, *Learned Optimism*.

12. Peterson, "The Future of Optimism."

13. Peterson et al., *Learned Helplessness*.

14. S. C. Segerstrom et al., "Optimism Is Associated with Mood, Coping, and Immune Change in Response to Stress," *Journal of Personality and Social Psychology* 74 (1998): 1646–55.

15. Monmaney, "'Have a Nice Day' May Translate to a Nicer Life."

16. For the best review of the relationship between anger and health, see R. Williams and V. Williams, *Anger Kills* (New York: Random House, 1993).

17. Ibid.

18. Ibid.

19. Brent Hafen et al., *Mind/Body Health* (Boston: Allyn and Bacon, 1996).

20. Ibid.

21. Robert Ornstein and David Sobel, *Healthy Pleasures* (Reading, Mass.: Addison-Wesley, 1989).

22. Williams and Williams, *Anger Kills*.

23. J. C. Barefoot et al., "Suspiciousness, Health, and Mortality: A Follow-Up Study of 500 Older Adults," *Psychosomatic Medicine* 49 (1987): 450–57.

24. E. V. Nunes et al., "Psychological Treatment for the Type A Behavior Pattern and for Coronary Artery Disease: A Meta-analysis of the Literature," *Psychosomatic Medicine* 48 (1987): 159–73.

25. M. W. Ketterer, "Secondary Prevention of Ischemic Heart Disease: The Case for Aggressive Behavioral Monitoring and Intervention," *Psychosomatics* 34 (1993): 478–84.

26. For a discussion of the effects of humor on the mind and body, see

David Sobel and Robert Ornstein, *The Healthy Mind, Healthy Body Handbook* (Los Altos, Calif.: DRx, 1996).

27. R. A. Martin and J. P. Dobbin, "Sense of Humor, Hassles, and Immunoglobulin A: Evidence for a Stress-Moderating Effect of Humor," *International Journal of Psychiatric Medicine* 18 (1988): 93–105.

28. P. Salovey et al., "Emotional States and Physical Health," *American Psychologist* 55 (2000): 110–21.

29. For an in-depth review of positive illusions, see Shelley Taylor, *Positive Illusions* (New York: Basic Books, 1989).

30. Ibid.

31. W. B. Cannon, "'Voodoo' Death," *American Anthropologist* 44 (1942): 169–81.

32. For a comprehensive review of the placebo effect, see H. Beecher, "The Powerful Placebo," *Journal of the American Medical Association* 159 (1955): 1602–6. Also, H. Benson, "The Placebo Effect," *Journal of the American Medical Association* 232 (1975): 1225–27; and A. K. Shapiro, "Placebo Effects in Medicine, Psychotherapy, and Psychoanalysis." In A. Bergin and S. Garfield, eds., *Handbook of Psychotherapy and Behavior Change* (New York: Wiley and Sons, 1971).

33. Beecher, "The Powerful Placebo."

34. S. Wolf, "Effects of Suggestion and Conditioning on the Action of Chemical Agents in Human Subjects: The Pharmacology of Placebos," *Journal of Clinical Investigation* 29 (1950): 100–109.

35. R. Lazarus, "The Costs and Benefits of Denial," in S. Benitz, ed., *Denial of Stress* (New York: International Universities Press, 1983).

36. Jean Piaget, *The Child's Conception of the World* (New York: Littlefield Adams, 1990).

37. Harold Koenig, Michael McCullough, and David Larson, *Handbook of Religion and Health* (New York: Oxford University Press, 2001).

38. D. Myers, "The Funds, Friends, and Faith of Happy People," *American Psychologist* 55 (2000): 56–57.

39. J. W. Yates et al., "Religion in Patients with Advanced Cancer," *Medical and Pediatric Oncology* 9 (1981): 121–28.

40. Koenig et al., *Handbook of Religion and Health*.

41. Herbert Benson, *Timeless Healing* (New York: Scribner, 1996).

Chapter 8: Social Support and Stress Hardiness

1. S. Maddi and S. Kobasa, *The Hardy Executive: Health Under Stress* (Chicago: Dorsey Professional Books, 1984).

2. J. W. Pennebaker et al., "Lack of Control as a Determinant of Perceived Physical Symptoms," *Journal of Personality and Social Psychology* 35 (1977): 167–74.

3. U. Lundberg and M. Frankenhaeser, "Psychophysiological Reactions to Noise as Modified by Personal Control over Stimulus Intensity," *Biological Psychology* 6 (1978): 51–58.

4. M. E. P. Seligman, *Learned Optimism* (New York: Knopf, 1991).

5. Ibid.

6. K. C. Corley et al., "Cardiac Responses Associated with 'Yoked Chair' Shock Avoidance in Squirrel Monkeys," *Psychophysiology* 12 (1975): 439–44.

7. E. J. Langer and J. Rodin, "The Effects of Choice and Enhanced Personal Responsibility: A Field Experiment in an Institutional Setting," *Journal of Personality and Social Psychology* 34 (1976): 191–98.

8. Allan Luks, *The Healing Power of Doing Good: The Health and Spiritual Benefits of Helping Others* (New York: Fawcett Columbine, 1991).

9. G. Vaillant, "Adoptive Mental Mechanisms: Their Role in a Positive Psychology," *American Psychologist* 55 (2000): 89–98.

10. J. House et al., "The Association of Social Relationships and Activities with Mortality," *American Journal of Epidemiology* 116 (1982): 123–40.

11. David Sobel and Robert Ornstein, *The Healing Brain* (New York: Touchstone, 1987).

12. One in five Americans changes residences every year and almost half relocate within any five-year period. See B. Hafen et al., *Mind/Body Health* (Boston: Allyn and Bacon, 1996).

13. The number of people living alone in the United States rose almost 400 percent between 1950 and 1980, the divorce rate has risen significantly in the past few years, and there has been an increasing trend toward staying unmarried. Ibid.

14. For a review of the effects of social support on health, see Sobel and Ornstein, *The Healing Brain*.

15. P. Salovey et al., "Emotional States and Physical Health," *American Psychologist* 55 (2000): 110–21.

16. J. House et al., "Social Relationships and Health," *Science,* July 29, 1988.

17. H. Koenig, M. McCullough, and D. Larson, *Handbook of Religion and Health* (New York: Oxford University Press, 2001).

18. D. Spiegal et al., "Effect of Psychosocial Treatment on Survival of Patients with Metastatic Breast Cancer," *Lancet* 2 (1989): 888–91.

19. S. Cohen et al., "Social Ties and Susceptibility to the Common Cold," *Journal of the American Medical Association,* June 1997, 1940–45.

20. M. A. Davis, "Living Arrangements and Survival Among Middle-Aged and Older Adults in the NHANES I Epidemiologic Follow-Up Study," *American Journal of Public Health,* March 1992, 401–6.

21. W. Ruberman et al., "Psychosocial Influences on Mortality After Myocardial Infarction," *New England Journal of Medicine* 311 (1984): 552–59.

22. J. K. Kiecolt-Glaser et al., "Psychosocial Modifiers of Immunocompetence in Medical Students," *Psychosomatic Medicine* 46 (1984): 7–14.

23. House et al., "Social Relationships and Health."

24. Ibid.

25. J. Kiecolt-Glaser et al., "Negative Behavior During Marital Conflict as Associated with Immunological Down-Regulation," *Psychosomatic Medicine* 55 (1993): 395–409.

26. J. Lynch, *The Broken Heart: The Medical Consequences of Loneliness* (New York: Basic Books, 1977).

27. A. Kraus and R. Lilienfeld, "Some Epidemiologic Aspects of the High Mortality Rate in the Young Widowed Group," *Journal of Chronic Diseases* 10 (1959): 207–17.

28. M. G. Marmot et al., "Epidemiologic Studies of Heart Disease and Stroke in Japanese Men Living in Japan, Hawaii, and California: Prevalence of Coronary and Hypertensive Heart Disease and Associated Risk Factors," *American Journal of Epidemiology* 102 (1975): 514–25.

29. James Pennebaker, *Opening Up: The Healing Power of Confiding in Others* (New York: William Morrow, 1990).

30. Sobel and Ornstein, *The Healing Brain.*

Chapter 9: Opening the Door to the AM: The Relaxation Response

1. See T. Wehr, "A Clock for All Seasons," in R. Buijs et al., *Progress in Brain Research* 11 (1996): 321–41.

2. See I. Kutz, J. Borysenko, and H. Benson, "Meditation and Psychotherapy: A Rationale for the Integration of Dynamic Psychotherapy, the Relaxation Response, and Mindfulness Meditation," *American Journal of Psychiatry* 142 (1985): 1–8. Primary-process thinking is a more receptive mode of mental functioning that employs feelings, images, and intuition. The flexibility during primary-process thinking involves a nonlinear integration of information that "connects" it more intuitively. The receptivity that characterizes primary-process thinking makes unconscious emotional stimuli more accessible. States of deep relaxation may energize us because they allow us to access and release unconscious emotional stimuli that have been causing unconscious activation of the stress response because the stimuli have been blocked from awareness. The release of unconscious emotional stimuli during the relaxation response often takes the form of images and feelings. This unusually receptive state is thought to represent the spontaneous emergence of unconscious feelings and emotions via images, symbols, and gestalts. See A. M. Green, "Brain Wave Training, Imagery, Creativity, and Integrative Experiments," *Proceedings of the Biofeedback Research Society* (Denver, Colo.: 1974).

3. I discussed the concept of releasing unconscious emotional information during states of deep relaxation in several lectures, including my keynote address at the 24th Annual Scientific Meeting of the Japanese Society of Autogenic Training, Tokyo, November 2001, entitled "Central Nervous System Mechanisms Mediating the Therapeutic Effects of the Relaxation Response and Autogenic Training"; and in several of my papers, including "The Effects of Short-Term Flotation Rest on Relaxation: A Controlled Study" (G. D. Jacobs, R. Heilbronner, and J. Stanley, *Health Psychology* 3 [1984]: 99–112) and "An Analysis of Cortical and Subcortical Processes Involved in Stress and Self-Regulation in the Human Nervous System from an Environmental Stimulation Perspective," for which I was awarded the John P. Zubek Memorial Award from the University of Manitoba and the International REST Investigators' Society in 1985. This concept was described earlier by several individuals, including K. Pelletier and C. Garfield (*Consciousness: East and West* [New York: Harper & Row, 1975]; Wolfgang Luthe, a German psychiatrist who studied the

clinical effects of a relaxation technique called autogenic training (W. Luthe, *Autogenic Training* [New York: Grune and Stratton, 1965]), and Louis West (cited in Daniel Schacter, "The Hypnagogic State: A Critical Review of the Literature," *Psychological Bulletin* 83 [1976]: 452–81).

Luthe described the neurophysiological changes during deep relaxation this way: "Brain-directed self-regulation mechanisms were actively engaging in the release of impulses from different parts of the brain; these discharges involve cortical, subcortical, and brain-stem mechanisms related to accumulated disturbed material." Luthe states that the goal of autogenic training is to "modify cortico-diencephalic interrelations and to allow the individual to attain a state of physiological quiescence in order to become more attuned to subcortical processes."

The crux of West's theory is that the reticular formation is responsible for integrating and inhibiting the vast amount of information impinging upon the brain and that inhibition requires energy. Under conditions of impaired sensory input, the reticular formation no longer receives the stimulation necessary to exert inhibitory effects on consciousness. As a result, sensory information normally inhibited is released.

4. Kutz, Borysenko, and Benson, "Meditation and Psychotherapy." Freud intended free association to be a technique that would enable patients' minds to relax to allow unconscious stimuli to enter the conscious mind. Freud intended free association to be pivotal to achieving catharsis—the relieving of anxiety by bringing "repressed" fears and feelings to consciousness.

5. T. Budzynski, "Biofeedback and Twilight States of Consciousness," in G. Schwartz and D. Shapiro, eds., *Consciousness and Self-Regulation*, vol. 1 (New York: Plenum, 1976).

6. For a comprehensive review of the hypnagogic state, see Schachter, "The Hypnagogic State." Schachter notes that the hypnagogic state has been described as "a thought, an intelligence, working within our own organization distinct from that of our own personality"; as a state in which "usually unheard things come to consciousness"; and, as a "dialogue between conscious and unconscious processes."

7. Sleep researchers do not consider stage 1 to be true sleep, for we are

easily awakened from it and have not lost consciousness of the external environment. In fact, if awakened from stage 1, most of us would maintain that we were not actually sleeping but that we were losing a direct train of thought and were experiencing hypnagogic imagery. Stage 1 produces a significant decrease in arousal in the AM as evidenced by decreases in the activity of the reticular formation and thalamus.

8. The free play of feelings, ideas, and images during reverie is considered to be of great importance in intuitively grasping the solution to a problem. Because creative insights can result from intuition and hunches, reverie may be a method of solving problems or achieving insights through the AM's intuitive processes. And because we have not been taught to be conscious of hypnagogic imagery, we don't notice it during states of reverie. It is also difficult to remember hypnagogic imagery if it is not recorded immediately after it occurs. However, studies have shown that people can be trained to observe their own hypnagogic imagery by being wakened when they enter reverie and asked to report or record the imagery on the spot. The ability to become aware of hypnagogic imagery depends in part on maintaining enough arousal to permit wakefulness and not fall asleep. If you trained yourself to become more conscious of the imagery that occurred during this state of deep relaxation, you would report emotionally laden images and symbols. And if you experience reverie often, you may be more likely to solve a problem or generate a new idea as a result of this state.

9. See H. Benson and M. Klipper, *The Relaxation Response* (New York: Avon, 1976).

10. Psychologist Robert Levenson uses the concept of "physiological arousal in excess of metabolic demand" to explain the deleterious effects of psychological stress. Levenson explains that, when people are running, a metabolic demand is created that is appropriate for the level of physiological arousal created by running. If the same level of physiological activity were produced when the person was sitting still, then arousal would be in excess of metabolic demand. Similarly, when we respond to stress with fight or flight, the metabolic demand is appropriate for the level of arousal. If the stress response is not accompanied by actual fighting or fleeing, then meta-

bolic demand cannot accommodate the physiological arousal and damage effects result from physiological pressure that is built up and not released. This pressure strains the mind and body, resulting in psychological and physiological damage. Levenson uses the analogy of a garden hose. No matter how high the water pressure, the hose will function without damage as long as the water is removed as powerfully as it is pumped in. But if the water is not released from the hose while it is running, pressure will build up and eventually damage the hose. See R. W. Levenson, "Emotional Control: Variation and Consequences," in *The Nature of Emotion,* ed. Paul Ekman and Richard Davidson (New York: Oxford University Press, 1994).

Psychologist Robert Thayer likens the process of "physiological arousal in excess of metabolic demand" to anaerobic energy metabolism. He suggests that stress and negative emotions like anxiety and nervousness, which are characterized by muscle tension with an absence of motor activity and also by shallow, uneven respiration, may cause physiological arousal through a process called anaerobic energy metabolism. Anaerobic metabolism, which breaks down energy stored in tissue as quickly as possible to supply the muscles, is also characterized by high muscle tension and results in a rapid increase in the production of lactic acid and some reduction in energy supplies. Because anaerobic energy metabolism is a very inefficient energy supply system, it results in fatigue. The increases in lactic acid may increase blood lactate levels, which may be associated with anxiety.

11. Biofeedback has been used to treat a variety of symptoms including headaches, anxiety, Raynaud's disease, hypertension, bruxism, asthma, chronic pain, insomnia, incontinence, and muscle disorders. A nice summary of the clinical and research applications of biofeedback can be found in the chapter entitled "Biofeedback: Using the Body's Signals" by Mark and Nancy Schwartz in *Mind/Body Medicine,* ed. Daniel Goleman and Joel Gurin (New York: Consumer Reports Books, 1993).

12. For a review of the therapeutic effects of the relaxation response, see H. Benson and E. Stuart, eds., *The Wellness Book* (New York: Fireside, 1992).

13. G. D. Jacobs, H. Benson, and R. Friedman, "Topographic EEG Map-

ping of the Relaxation Response," *Biofeedback and Self-Regulation* 21 (1996): 121–29.

14. For a summary of the EEG changes associated with relaxation response techniques, see M. A. West, "Meditation and the EEG," *Psychological Medicine* 10 (1980): 369–75.

15. G. D. Jacobs and J. F. Lubar, "Spectral Analysis of the Central Nervous System Effects of the Relaxation Response Elicited by Autogenic Training," *Behavioral Medicine* 15 (1989): 125–32.

16. Several researchers (see West, "Meditation and the EEG") have suggested that the RR may be similar to stage 1 sleep. They concluded that, because alpha and theta EEG occur during RR, and alpha and theta are the primary EEG patterns prior to and during stage 1 sleep, the brain wave changes during the relaxation response are consistent with quiet wakefulness, the onset of stage 1 sleep, and the hypnagogic state—the phenomenological transition between waking and sleeping. In fact, sleep researchers who score the EEGs of people practicing the RR note a high percentage of stage 1 sleep. Thus, deep relaxation during the RR may in fact be a finely held stage 1 sleep state. However, during the RR, the state is consciously induced and held without falling into actual sleep. In this sense, the RR may represent a suspended or "frozen" state between waking and sleeping. And since the RR may be similar to stage 1 sleep, it is probable that, like stage 1 sleep, the RR decreases arousal in the AM by reducing activity in the reticular formation and thalamus, which would be consistent with a deep state of relaxation in the brain.

17. For an in-depth discussion of the neurophysiological effects of meditation, see Robert Ornstein's classic book *The Psychology of Consciousness* (San Francisco: W. H. Freeman, 1972) and Kenneth Pelletier and Charles Garfield's *Consciousness: East and West* (New York: Harper and Row, 1975).

18. Conclusive evidence that the human sleep/wake rhythm is characterized by a strong propensity for a mid-afternoon nap was first demonstrated by Dr. Scott Campbell in 1986. He found that human sleep is biphasic, consisting of a long nocturnal period of sleep and a brief (one- to two-hour) afternoon sleep period that fell about twelve hours from the middle of the nocturnal sleep period. Other evidence for the afternoon pressure for a nap includes the fact that sleepiness

increases in the afternoon. The afternoon increase in sleepiness coincides with an increase in sleepiness-related accidents. In fact, deaths from all causes show a secondary peak in the afternoon after a nocturnal peak, presumably from sleepiness-related accidents. For a review of the scientific findings concerning napping, see D. Dinges and R. Broughton, eds., *Sleep and Alertness: Chronobiological, Behavioral, and Medical Aspects of Napping* (New York: Raven Press, 1989).

19. T. Wang et al., "Responses of Natural Killer Cell Activity to Acute Laboratory Stressors in Healthy Men at Different Times of Day," *Health Psychology* 17 (1998): 428–35.

20. Relative to waking and dream sleep, brain imaging studies of deep sleep show a decrease in global cerebral energy metabolism, global cerebral blood flow, and blood flow velocity which, at least for energy metabolism, decreases progressively with greater depth of sleep. Deactivation during deep sleep is seen in the brain stem and reticular formation, the thalamus and limbic system, and the prefrontal cortex. During dream sleep, global cerebral energy metabolism tends to be equal to, or greater than, that of waking. For a review of the neural substrates of sleep, see P. Maquet, "Functional Neuroimaging of Normal Human Sleep by Positron Emission Tomography," *Journal of Sleep Research* 9 (2000): 207–31; and J. A. Hobson, E. F. Pace-Schott, and R. Stickgold, "Consciousness: Its Vicissitudes in Waking and Sleep," in *The New Cognitive Neurosciences*, ed. M. Gazzaniga (Cambridge, Mass.: Bradford Press, 2000).

21. Hobson, Pace-Schott, and Stickgold, "Consciousness: Its Vicissitudes in Waking and Sleep."

22. In my research, I found that, compared to good sleepers, insomniacs exhibit faster brain waves before and during the early stages of sleep. See G. D. Jacobs, H. Benson, and R. Friedman, "Home-Based Central Nervous Assessment of a Multifactor Behavioral Intervention for Chronic Sleep-Onset Insomnia," *Behavior Therapy* 24 (1993): 159–74. Other researchers have also shown that insomniacs exhibit faster brain waves during dream sleep. Taken together, these findings suggest that insomniacs have a more difficult time sleeping, in part, because their brain is too activated. For a more detailed discussion of insomnia, the effects of stress on sleep, and an empirically validated

drug-free program for managing insomnia, see my book *Say Good Night to Insomnia* (New York: Henry Holt, 1999).

Chapter 10: Before Words, Images Were

1. H. Benson et al., "Three Case Reports of the Metabolic and EEG Changes During Advanced Buddhist Meditation Techniques," *Behavioral Medicine* (Summer 1990): 1690–95.

2. A 1936 study found that images such as being attacked by a leopard caused greater increases in heart rate, respiration, and skin sweating than images such as a flock of grazing sheep. For a review of this study and others involving emotional imagery, see P. Lang, "A Bio-Informational Theory of Emotional Imagery," *Psychophysiology* 16 (1979): 495–512. Also, see E. Di Giusto and N. Bond, "Imagery and the Autonomic Nervous System: Some Methodological Issues," *Perceptual and Motor Skills* 48 (1979): 427–38.

3. Edmund Jacobsen, *Progressive Relaxation* (Chicago: University of Chicago Press, 1938).

4. Researchers studied the physiological responses of forty-six females who were asked to imagine fearful and neutral scenes. Each subject composed two fifty-word scripts with the experimenters based on both a personally frightening and a neutral topic. They were first told to listen to the script while it was read to them; then, to visualize the script. There were no physiological changes when the script was read. However, when the script was visualized, heart rate and skin conductance increased significantly. See Lang, "A Bio-Informational Theory of Emotional Imagery."

5. G. E. Schwartz, L. Ahern, and S. L. Brown, "Lateralized Facial Muscle Response to Positive and Negative Emotional Stimuli," *Psychophysiology* 16 (1979): 561–71.

6. Lang and his colleagues studied spider phobics and found greater stress responsivity during imagined fearful scenes compared to neutral scenes. Self-report ratings of fear about various images correlated highly with changes in heart rate. See Lang, "A Bio-Informational Theory of Emotional Imagery."

7. S. L. Rauch et al., "A Symptom Provocation Study of Post-Traumatic Stress Disorder Using Positron Emission Tomography and Script-

Driven Imagery," *Archives of General Psychiatry* 53 (1996): 380–87. Also, R. Bryant and A. Harvey, "Visual Imagery in Posttraumatic Stress Disorder," *Journal of Trauma and Stress* 9 (1996): 613–19.

8. Joseph Wolpe, *Psychotherapy by Reciprocal Inhibition* (Stanford, Calif.: Stanford University Press, 1958).

9. For a review of this study and other experimental findings on imagery, see M. Rossman, *Healing Yourself: A Step-By-Step Program for Better Health Through Imagery* (New York: Pocket Books, 1987).

10. K. Syrjala et al., "Relaxation and Imagery and Cognitive-Behavioral Training Reduce Pain During Cancer Treatment: A Controlled Clinical Trial," *Pain* 63 (1995): 189–98.

11. B. Gruber et al., "Immune System and Psychological Changes in Metastatic Breast Cancer Patients Using Relaxation and Guided Imagery: A Pilot Study," *Scandinavian Journal of Behavior Therapy* 17 (1988): 25–46.

12. See L. Gottschalk et al., "The Effects of Anxiety and Hostility in Silent Mentation on Localized Cerebral Glucose Metabolism," *Comprehensive Psychiatry* 33 (1992): 52–59.

13. S. Kosslyn and W. Thompson, "Shared Mechanisms in Visual Imagery and Visual Perception: Insights from Cognitive Neuroscience," in *The New Cognitive Neurosciences*, ed. M. Gazzaniga (Cambridge, Mass.: MIT Press, 2000).

Chapter 11: The Ancestral Mind's Minimal Daily Requirements

1. For a review of the latest scientific findings on biology and music, see P. Gray et al., "The Music of Nature and the Nature of Music," *Science* 291 (2001): 52–56.

2. Howard Gardner, *Frames of Mind: The Theory of Multiple Intelligences* (New York: Basic Books, 1993).

3. Gray et al., "The Music of Nature and the Nature of Music."

4. Ibid.

5. Ibid.

6. G. Cook, "Wired for Sound," *The Boston Globe*, April 15, 2001, 6–7.

7. See David Chandler, "Ancient Note: Music as a Bridge Between the Species," *The Boston Globe*, January 5, 2001.

8. R. Thayer, *The Origin of Everyday Moods* (New York: Oxford Univer-

sity Press, 1996). Thayer has conducted some of the best research on moods and their management.

9. Robert Ornstein and David Sobel, *Healthy Pleasures* (Reading, Mass.: Addison Wesley, 1989).

10. David Sobel and Robert Ornstein, *The Healthy Mind, Healthy Body Handbook* (Los Altos, Calif.: DRx, 1996).

11. Ibid.

12. Ibid.

13. Ibid.

14. P. Gray et al., "The Music of Nature and the Nature of Music."

15. A. Goldstein, "Thrills in Response to Music and Other Stimuli," *Physiological Psychology* 8 (1980): 126–29.

16. I presented my preliminary findings on the effects of the relaxation response and music on EEG activity at the conference titled "Science and Mind/Body Medicine" held by the Mind/Body Medical Institute and Harvard Medical School in Boston, May 3–5, 2001.

17. For a review of the scientific findings on bright light, see R. Lam, ed., *Seasonal Affective Disorder and Beyond: Light Treatment for SAD and Non-SAD Conditions* (Washington, D.C.: American Psychiatric Press, 1998).

18. A. Wirz-Justice, "Beginning to See the Light," *Archives of General Psychiatry* 55 (1998): 861–62. The seasonal changes in mood observed in SAD patients may represent the extreme end of a behavioral spectrum, which can also be observed in normal, nondepressed individuals to a lesser extent.

19. S. Kasper et al., "Epidemiological Findings of Seasonal Changes in Mood and Behavior," *Archives of General Psychiatry* 46 (1989): 823–27.

20. Lam, *Seasonal Affective Disorder and Beyond.*

21. Wirz-Justice, "Beginning to See the Light."

22. T. Wehr, "A Clock for All Seasons," in R. Buijs et al., *Progress in Brain Research* 11 (1966).

23. Ibid. Although we have retained these intrinsic seasonal responses to changes in light and dark, the responses are now masked by indoor environments, control over lighting, shift work, and air travel. Instead of being exposed to purely natural light, we expose ourselves to artificial indoor light and virtually eliminate our exposure to true

darkness as a consequence of nighttime lighting. This disruption in the effect of light on the circadian pacemaker is believed to be responsible for the increased predisposition to seasonal changes in mood in vulnerable individuals, possibly through disruptions in the balance of neurotransmitters like serotonin that are involved in mood.

24. For a discussion of the relationship between melatonin, sleep, and body temperature, see Gregg Jacobs, *Say Good Night to Insomnia* (New York: Henry Holt, 1999).

25. J. Brody, "Persuading Potatoes to Get Off the Couches," *The New York Times*, February 2, 1999.

26. K. Cullen, "The Childhood Obesity Epidemic," *The Boston Globe*, October 3, 2000.

27. K. Cullen, "Obesity Study Tells English: Hold the Burgers and Chips," *The Boston Globe*, February 16, 2001.

28. I. Lakshmana, "Rise in Obesity Weighs on Asia," *The Boston Globe*, February 27, 2001.

29. Ibid.

30. Cullen, "The Childhood Obesity Epidemic."

31. Ibid.

32. Ibid.

33. Brody, "Persuading Potatoes to Get Off the Couches."

34. Ibid.

35. Ibid.

36. F. Booth et al., "Waging War in Modern Chronic Diseases: Primary Prevention Through Exercise Biology," *Journal of Applied Physiology* 88 (2000): 774–87.

37. For a review of the many benefits of physical activity, see P. Fentem, "Exercise in Disease Prevention," *British Medical Bulletin* 48 (1992): 630–50. If we would adopt a physically active lifestyle, we could prevent most chronic diseases, positively impact all known chronic diseases, decrease morbidity while increasing longevity and vitality, and reduce health care costs by billions.

38. Brody, "Persuading Potatoes to Get Off the Couches."

39. Thayer, *The Origin of Everyday Moods*.

40. Ibid.

41. Ibid.

42. David Sobel and Robert Ornstein, *The Healthy Mind, Healthy Body Handbook.*

43. M. Babyak et al., "Exercise Treatment for Major Depression: Maintenance of Therapeutic Benefit at 10 Months," *Psychsomatic Medicine* 62 (2000): 633–38.

44. D. Kong, "Exercise Seen Boosting Children's Brain Function," *The Boston Globe*, November 9, 1999.

45. Jacobs, *Say Good Night to Insomnia.*

46. Ibid.

47. I-Min Lee et al., "Physical Activity and Coronary Heart Disease in Women," *Journal of the American Medical Association* 285 (2001): 1447–54.

Chapter 12: Solitude and Wilderness: Medicine for the Soul

1. Anthony Storr, in *Solitude: A Return to the Self* (New York: Ballantine, 1988), notes that modern psychotherapists have used as their criteria of emotional maturity the capacity of the individual to "make mature relationships on equal terms" and to form meaningful, secure attachments with others. Because man is a social being, it is widely believed that intimate relationships are the primary source of human happiness. But Storr points out that many creative individuals like Descartes, Newton, and Kant did not form close personal ties and that solitude may have been an important component of their creative talent. With few exceptions, psychotherapists have failed to consider that the capacity to be alone is also an aspect of emotional maturity. Not only is the capacity to be alone one component of an inner security that has built up over years; Storr also believes that solitude is adaptive and necessary for optimal brain functioning. Because some children who enjoy the "solitary exercise of the imagination" may develop creative potential, Storr believes that we should ensure that our children are given time and opportunity for solitude.

2. Various relaxation techniques that elicit the relaxation response specifically incorporate solitude and sensory restriction as a means of enhancing relaxation. Autogenic training is facilitated by conditions that involve a significant reduction of incoming and outgoing sensory signals. In this sense, autogenic training can be understood as a technique of self-induced sensory deprivation. Biofeedback usu-

ally occurs in an environment of moderately reduced stimulation, and progressive muscle relaxation is ideally learned in a soundproof room, or, if this is not available, an environment of reduced auditory and visual stimulation. For a review of the use of sensory restriction and the use of a variety of relaxation techniques, see Peter Suedfeld, *Restricted Environmental Stimulation: Research and Clinical Applications* (New York: Wiley and Sons, 1980).

3. See G. D. Jacobs, R. Heilbronner, and J. Stanley, "The Effects of Short-Term Flotation Rest on Relaxation: A Controlled Study," *Health Psychology* 3 (1984): 99–112. Much of the earlier scientific literature on "sensory deprivation" reported negative effects: boredom, anxiety, or even hallucinations. However, these findings were not replicated, and later studies demonstrated that these earlier findings were significantly biased by experimental demands (uncomfortable settings, negative expectations on the part of the experimenters that were conveyed to subjects, and the like). Later studies that created more comfortable sensory restriction environments and that were careful not to induce negative expectations reported that sensory deprivation could be positive and therapeutic. Peter Suedfeld, the world's authority on what later came to be known as REST (restricted environmental stimulation therapy), has written at length about the benefits of solitude and isolation. His research and other studies have demonstrated convincingly that solitude and environments of reduced sensory stimulation produce therapeutic effects that are similar to those of meditation and other relaxation therapies. See Suedfeld, *Restricted Environmental Stimulation*.

4. Suedfeld points out that Merton is probably the most lyrical of the proponents of solitude. Merton draws many parallels between the early desert fathers and the European and Oriental monks and emphasizes the role of solitude as a contributor to the religious life. Religious retreats for the layperson and the clergy incorporate periods of solitude for the same reasons.

5. Aaron Katcher et al., "Looking, Talking and Blood Pressure: The Physiological Consequences of Interaction with the Living Environment," in A. H. Katcher and A. M. Beck, eds., *New Perspectives on Our Lives with Animal Companions* (Philadelphia: University of Pennsylvania Press, 1983). Katcher believes that viewing nature relaxes us

because nature draws attention outward, interrupts the flow of thoughts, and produces states of reverie. This reverie state is similar to the relaxation that one feels during the relaxation response, when watching the ocean, or sitting in front of a fire. Like many stimuli in nature, a fire is always different yet always the same. The flames change form continually, the logs burn and glow, yet one fire is indistinguishable from the next. The beauty of the fire captures attention; the sudden flash of flame, the glowing of the coals offer constant novelty yet are always the same. The combination of beauty and monotony, novelty and constancy attracts our attention and elicits deep states of tranquillity in the AM that reenergize and renew us.

6. R. S. Ulrich, "View Through a Window May Influence Recovery from Surgery," *Science* 224 (1984): 420–21.

7. For a discussion of evolved responses to landscapes, see G. H. Orians and J. H. Heerwagen, "Evolved Responses to Landscapes" in *The Adapted Mind*, ed. J. H. Barkow, L. Cosmides, and J. Tooby (New York: Oxford University Press, 1998).

8. Ibid.

9. According to evolutionary psychologists Gordon Orians and Judith Heerwagen, habitat selection involves emotional responses to key features of the environment. These features induce the positive and negative emotions that determine whether an organism will reject, explore, or settle in a habitat. The ability of a habitat to evoke positive emotional states would evolve to be positively correlated with expected survival and reproductive success of an organism. See ibid.

10. See the Associated Press article "Extinction Risk Mounting" by Mara Bellab, *The Boston Globe*, September 29, 2000.

Acknowledgments

I want to thank Bruce Hetzler and John Stanley for encouraging my study of the Ancestral Mind while I attended Lawrence University. I also want to thank Allan Belden, who gave me my first opportunity to observe and teach the healing power of the Ancestral Mind, and Joel Lubar, who mentored me during my graduate work at the University of Tennessee. The work of several pioneers in the field of mind/body medicine, including Herbert Benson, Kenneth Pelletier, Robert Ornstein, and David Sobel, was instrumental in shaping my thinking about the Ancestral Mind. The groundbreaking work of Joseph LeDoux, Daniel Goleman, and Richard Davidson on emotions crystallized my thinking about the Ancestral Mind; for this book I drew significantly on their work. I owe much to my colleagues Allan Hobson, Ed Pace-Schott, and Bob Stickgold at the Laboratory of Neurophysiology at Harvard Medical School for our discussions and seminars that focused my thinking throughout the writing of this book. They are talented and dedicated scientists.

I also want to thank Dennis Russo and Bruce Masek for giving me the opportunity to learn about the power of the Ancestral Mind in children during my work at Children's Hospital in Boston. I am extremely grateful to Herbert Benson, who had a significant influence on my career. He supported my research in mind/body medicine over the years at Harvard Medical School and the Mind/Body Medical Institute through the Sherman-Warburg fellowship, the Dosberg Foundation, and Procter and Gamble. Thanks to two of my colleagues at the Mind/Body Medical Institute, Ann Webster and the late Richard Friedman, for sharing their wisdom, talents, and friendship. The Heart, Lung, and Blood Institute at the National Institutes of Health, as well as Lifewaves International, funded my research on insomnia.

I owe a great deal to my editor at Viking Penguin, Rick Kot. Rick challenged me, made me work hard throughout the revisions, and had the wisdom to keep saying it was not right until it was. He is a patient, smart, and talented editor and it was a privilege to work with him.

I also owe a lot to Bill Patrick, who was directly involved in the writing and revisions of this book and in delivering its message. He challenged me to sharpen my ideas, and his efforts enabled this book to evolve out of a scien-

232   Acknowledgments

tific voice and into an accessible one without compromising my ideas. It was exciting to work with an expert like Bill; this book would not have been possible without his knowledge and skill.

I am very fortunate to have Felicia Eth as my literary agent. She had the instincts to see this book's potential and the talent to make it a reality. She shaped my ideas and my proposal and provided unending time, creative input, and professional advice. Her patience and dependability are deeply appreciated.

My parents, Jim and Laverne, have given me constant support and encouragement; my dad also gave me feedback on the entire manuscript.

Finally, I am grateful to my wife, Jody Skiest, who reviewed and offered feedback on the manuscript and provided encouragement, support, and advice throughout every stage of this project.

Index